COUNTRY
WOOLWORK

COUNTRY
WOOLWORK

25 SIMPLE WOOL EMBROIDERY PROJECTS

Melinda Coss

Trafalgar Square Publishing

For
Stephen, Helen, Johnny and Joe Grant.

With long distance hugs.

First published in the United States of America in 1997
by Trafalgar Square Publishing,
North Pomfret, Vermont 05053

First published in Great Britain in 1997
by Collins & Brown Limited

Printed and bound in Singapore by C.S. Graphics

1 3 5 7 9 8 6 4 2

ISBN 1-57076-087-X

Library of Congress Catalog Card Number: 96-62081

Editor: Emma Callery
Designer: Roger Daniels
Photography: Jon Stewart

Reproduction by Classic Scan

Contents

Introduction

O F ALL OF NATURE'S GIFTS, wool has to be one of the most comforting. Over the centuries it has clothed us, kept us warm and decorated our homes. In babyhood it offers us security, and in adulthood the soft and sensual touch of pure wool has a unique and marvellous way of putting the world to rights.

In this book, wool is both a background fabric and an embroidery thread. It is the perfect medium for beginners as it provides speedy results that do not require the precise skills of cottons and silks. Within these pages you will find a library of basic stitches with which you can create flower gardens and country images with the minimum of time and effort. By grouping the stitches in different ways you can build up your own collection of flowers and, in the Flower Library on pages 114-115 you will find step-by-step instructions for creating some perennial favourites. The pages of motifs on pages 116-122 will give you ideas on how to arrange them. These motifs can be stitched on blanketing, knitted fabrics, calico and denim and used to decorate clothing, soft toys or soft furnishing items that you have either bought ready-made or sewn yourself. For those who are terrified of anything more complicated than a simple straight stitch, look at the chicks on pages

95-97 or the teddy bears on pages 98-100. These are created by outlining the shape on your fabric with a basic running stitch and filling in the detail with random straight stitches to represent fur and feathers.

Buttons, ribbons and thick cotton perlé threads also have their place in this book as they provide a simple way of highlighting a design. Look for interesting buttons at yard sales and jumble sales and never discard a piece of clothing without keeping the buttons for future embroidery projects. Even the humble shirt button can look wonderful stitched to your fabric with a contrasting knot to form the flower centre.

Small projects, such as box tops and pincushions can be worked in an evening and make delightful gifts. You can also monogram items using the floral alphabet on pages 75-79 to create a very personal gift.

If you are traditionally a cross stitcher, wool embroidery offers a creative and addictive alternative. You will be amazed at the different effects you can achieve and how quickly your work will develop. Above all, this is a chance to be truly creative with very basic stitches. Enjoy yourself.

MELINDA COSS

Beautiful Brights

Bold, beautiful and primitive designs can be created very
easily from a limited palette of brightly coloured
wools. In this section I have used felt and calico in addition
to wool blanketing as background fabrics, but you can
use any fabric that takes the weight of the wool.
Experiment with clashing colours and highlight your
designs with bright buttons and beads. Treat your needle
and thread as you would a coloured crayon and have
fun outlining and filling in shapes. This section is
designed for those who love to play with colour –
don't worry about the quality of your stitching,
it is the overall look that matters.

Poppy Blanket

Embroidery on woollen blanket fabric is a great idea I've picked up from Australia. Fun to do, it offers a practical use for your completed work and is perfect if you prefer working freestyle with wools on chunkier fabrics.

PREPARATION

Enlarge the charts overleaf by 125 per cent and then trace them onto tissue paper. Then transfer the design onto the blanketing by positioning the outlines as follows:

Motif A: centred on the blanketing.

Motif B: two of each on each edge of the blanket, one turned over to make a mirror image.

Motif C: one positioned towards each corner between motifs A and B.

With small running stitches and sewing thread, stitch through the tracings to form outlines. Tear away the tracing paper when every line has been covered.

STITCHING

Following the chart overleaf and its key, work the poppy petals in red (8200) and coral (8212) straight stitch, the poppy centres in dark brown (9666) colonial knots, the corn ears in autumn gold (8058 and 8060) detached chain stitch, the poppy leaves in green (9176) fly stitch and the daisies in white (8002) straight stitch, spanning out from a central point. Work the forget-me-nots in blue (8776) detached chain stitch and then add their centres and the daisy centres in pale yellow (8092) colonial knots. Add a few fly stitches in light green (9058) beside some of the forget-me-not flowers.

FINISHING

When the design is complete, fringe the edge of the blanket for 1 cm (½ in), pulling out the threads one at a time.

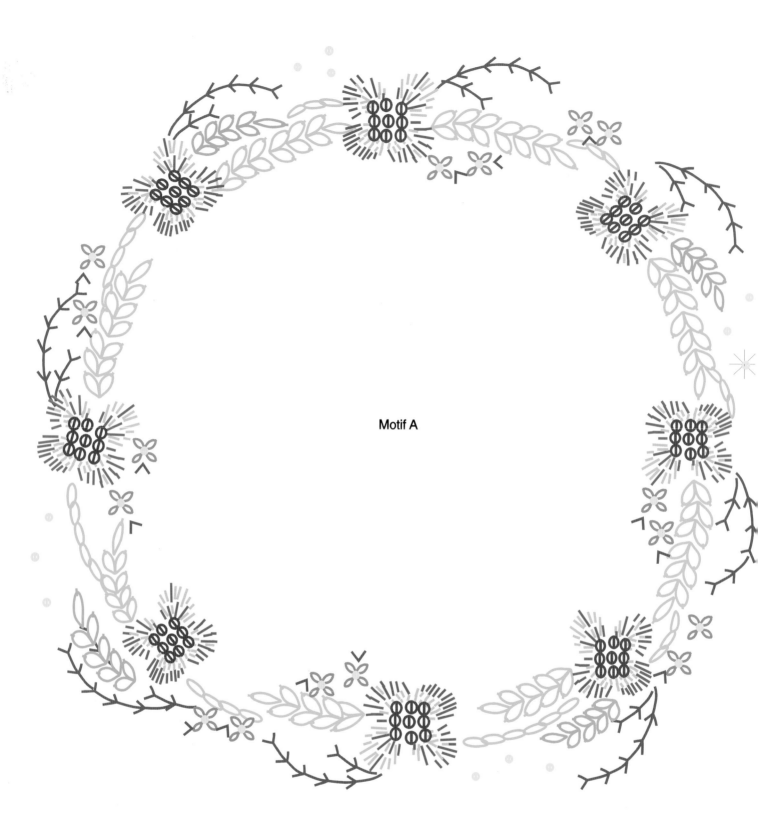

Motif A

Motif B

Motif C

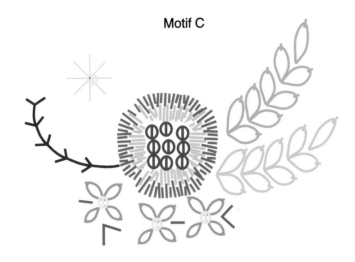

		detached chain stitch 8060
		detached chain stitch 8058
		detached chain stitch 8776 with **colonial knot** 8092
		fly stitch 9058
		colonial knot 9666
		straight stitch 8002 with **colonial knot** 8092
		fly stitch 9176
		straight stitch 8200 with 8212 centre
		stem stitch 8060
		stem stitch 8058

EMBROIDERY MATERIALS

STITCHES (SEE PAGES 110-113)

Colonial knot

Detached chain

Fly

Straight

MATERIALS

Pencil

Tracing paper

Transfer pencil or water-soluble pen

1 piece of pure wool blanketing 2.5 cm
(1 in) larger all around than the box top

Medium-sized crewel needle

1 piece of polyester wadding (batting)
cut to the same size as the box top

Rubber-based adhesive

1 piece of mounting card cut to the
same size as the box top

1 box (ours is 12.5 cm [5 in] diameter)

YARNS

Anchor Tapisserie Wool

1 skein of each of the
following colours:

Red 8200

Coral 8212

Green 9176

Blue 8776

White 8002

Gold 8058

Old gold 8060

Brown 9666

Yellow 8092

Light green 9058

Poppy Box

You can make a box top for any shape box using the following method, creating a gift very quickly. Here I have used one of the poppy and corn motifs featured on the blanket on pages 10-13. However, you may decide to use a decorative monogram (see pages 72-79) instead.

PREPARATION

Trace off the outline of the design and transfer this to the centre of the fabric either with the transfer pencil or by cutting out the basic shapes and drawing around them with the water-soluble pen.

STITCHING

Work the centre of the poppy, filling it with colonial knots in brown (9666). Then work the inside of the petals in coral (8212) straight stitch alternating the lengths of each stitch. Repeat around the outer edge of the flower using red (8200) wool. Draw guidelines on your fabric for the corn and work pairs of detached chain stitch for the ears and form the point with a single detached chain stitch. Work the green (9176) leaf in a curved row of fly stitches, then the forget-me-nots, grouping four blue (8776) detached chain stitches around a yellow (8092) colonial knot. Add fly stitches in light green (9058) for leaves. Work a single flower from crossed straight stitches in white (8002) with a yellow (8092) colonial knot at the centre.

MAKING UP

Glue the wadding (batting) to the card. Stretch the embroidery over the wadding (batting) and glue the edges to the underside of the card. Glue the covered card to the box top. Baste a twisted wool cord around the edge to cover the join. To make a twisted wool cord, open out a skein of wool and fold it into four lengths. Twist tightly, fold in half and let the two halves twist together. Secure the ends.

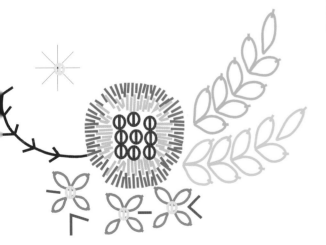

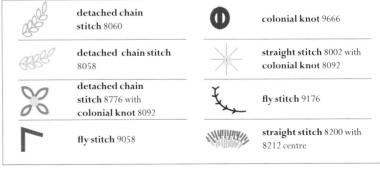

	detached chain stitch 8060		colonial knot 9666
	detached chain stitch 8058		straight stitch 8002 with colonial knot 8092
	detached chain stitch 8776 with colonial knot 8092		fly stitch 9176
	fly stitch 9058		straight stitch 8200 with 8212 centre

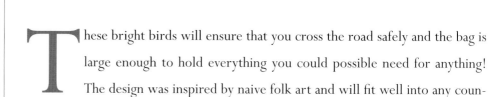

Bird Bag

These bright birds will ensure that you cross the road safely and the bag is large enough to hold everything you could possible need for anything! The design was inspired by naive folk art and will fit well into any country setting. If you don't want to make the bag from scratch you can stitch the motifs onto a ready-made shopping bag or use them as a panel on a plain cushion.

PREPARATION

Trace off the outlines of the bird and heart motifs overleaf (for the heart, follow the fuchsia [8456] chain stitch heart for the outline) and cut three birds and four hearts from the paper. Cut off 18 cm (7 in) from the end of the strip of blanketing for the two heart panels on the flap. Starting with the central bird, pin the cut outs in a row on the remaining blanketing, alternating the hearts and birds. Draw around the cut outs onto the fabric using the water-soluble pen.

STITCHING

Hearts Embroider over the outline with one row of fuchsia (8456) chain stitch. Then work an inner row of yellow (8096) chain stitch. Place evenly-spaced green (9116) colonial knots around the outer edge.

Birds Work the bodies in blue (8644) chain stitch, spiralling in to the centre. Work the heads in a series of half circles of chain stitch beginning at the outer edge and working inwards and following the colours on the chart. Add a yellow (8096) colonial knot for each eye and beak in satin stitch. Join the heads and bodies with rows of zigzag straight stitch following the chart overleaf for the choices of colour. Each foot is made up from one fly stitch in royal blue (8692). For the tail feathers, draw four guidelines onto the fabric and work rows of fly stitch on the top, again following the chart for the colours. Take the left-over blanketing and cut it in half. Embroider a heart on each piece, as for the main panel.

MAKING UP

Cut four pieces from the hessian (canvas) as follows:

Front top panel 43 x 26 cm (17 x 10½ in)

Front bottom panel 43 x 13 cm (17 x 5½ in)

Back panel (and flap) 43 x 86 cm (17 x 34 in)

Handle 102 x 10 cm (40 x 4 in)

Take the front top hessian (canvas) panel and make a 1 cm (½ in) double hem along the top (long) edge. Place the panel right sides down on the embroidered panel aligning the top of the embroidery with the bottom of the hessian (canvas). Stitch together, taking a 1 cm (½ in) seam allowance. Open out and sew the bottom edge of the embroidery to the top of the front bottom hessian (canvas) panel in the same way with right sides together. Fold under and pin a 1 cm (½ in) hem on each side and along the bottom of this panel and press firmly with a steam iron. Take the back piece of fabric and pin a 1 cm (½ in) hem all round the outer edge and again press firmly.

To make the flap, lay the front piece, wrong sides together, on top of the back piece aligning the bottom edges. Fold down the top edge of the back piece so that it lies just beneath the top of the front panel. Remove the front, pin the flap in place on the back, press the top crease and stitch the side seams to create a double thickness flap. Pin the front over the back, positioned as before, and machine stitch the sides and bottom of the bag on the right side of the fabric on the hessian (canvas) only. Slipstitch the sides of the woolwork panel in place.

To make the handle, take the remaining strip of hessian (canvas) and turn up the short edges by 1 cm (½ in) and press. Turn in and press the long edges in the same way. Fold the strip in half lengthwise and, on the right side of the fabric, machine or hand sew the seam close to the edge of the fabric. Top stitch the fold in the same manner to make a nice flat handle. Stitch the ends of the handle in place on the top side seams of the bag.

Take the two heart panels and cut them into evenly-sized oblong panels measuring approximately 11.5 x 8 cm (4½ x 3¼ in). Pin them into position on the bag flap 7.5 cm (3 in) in from the outside edges and 2.5 cm (1 in) up from the bottom of the flap. Hand stitch these in place with white thread, tucking under the raw edges of blanketing as you sew.

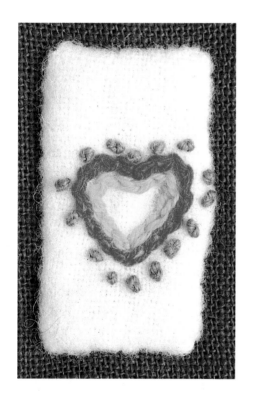

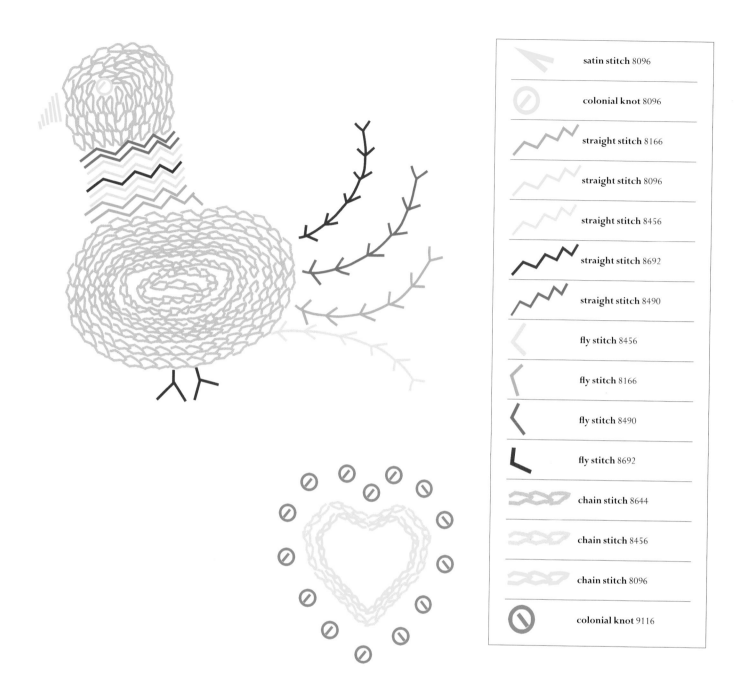

satin stitch 8096
colonial knot 8096
straight stitch 8166
straight stitch 8096
straight stitch 8456
straight stitch 8692
straight stitch 8490
fly stitch 8456
fly stitch 8166
fly stitch 8490
fly stitch 8692
chain stitch 8644
chain stitch 8456
chain stitch 8096
colonial knot 9116

Dandelion Hamper

By mounting your embroidery on a piece of card you can create a panel that can be used in numerous decorative ways. Here I have stitched the panel to the top of a wicker hamper, but you could also stick the panel to the front of a book, a box or a cupboard door – or anything else that needs jazzing up. The dandelion leaves are stitched in pairs of buttonhole stitches (see blanket stitch on page 110) with the tops angled so they join to form a triangle. The official name for this stitch is closed buttonhole stitch.

PREPARATION

Using the pencil, trace the design from the chart overleaf. Position the tracing centrally on the blanketing and mark the positions of the flowers through the paper with the water-soluble pen.

STITCHING

Begin with the dandelion heads, working straight stitches in the directions and colours indicated on the chart. Do not worry if the stitches overlap as this will give you a three-dimensional effect. Also, work the dandelion buds in straight stitches as outlined on the chart. Next, work the dandelion seed heads in shades of stone (9384) and cream (9506) fly stitches, as indicated on the chart. Work all the stems in stone (9384) stem stitch, then add the light green (9164) straight stitches under the flower heads.

Now work the sprays of leaves which consist of two back-to-back rows of closed buttonhole stitch in shades of green (9168 and 9164). The forget-me-nots are groups of four detached chain stitches in blue (8774), finished in the centre with a light yellow (8016) colonial knot. Finally, work the random colonial knots and fly stitches and the single lazy daisy stitch following the chart for the colours.

MAKING UP

Lightly glue the wadding (batting) to the card and lay the embroidery face down on the table. Place the covered card, wadding (batting) side down over the embroidery making sure that the design is properly centred. Turn the waste edges to the back of the card and either glue these down or lace them into place. Take care to tuck in the corners neatly to avoid bulk. Once the glue is dry, using sewing thread carefully stitch the panel to the top of the hamper around the underside of the edges. Or stick the panel to the top of a box, or whatever you are covering.

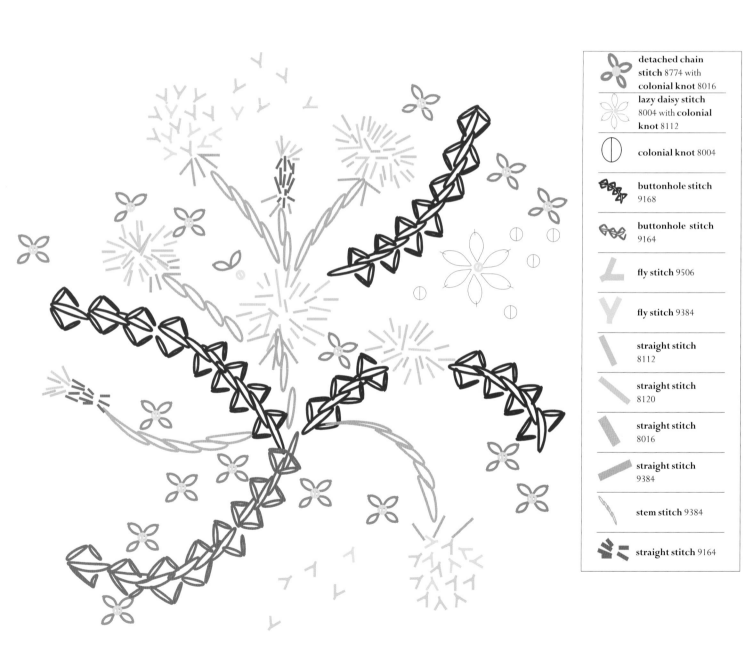

	detached chain stitch 8774 with **colonial knot** 8016
	lazy daisy stitch 8004 with **colonial knot** 8112
	colonial knot 8004
	buttonhole stitch 9168
	buttonhole stitch 9164
	fly stitch 9506
	fly stitch 9384
	straight stitch 8112
	straight stitch 8120
	straight stitch 8016
	straight stitch 9384
	stem stitch 9384
	straight stitch 9164

Cow Bag

While most of this book is devoted to embroidering woollen fabric with wool, you can, of course, work on any medium- to heavy weight fabric. Calico, for example, is a cheap and effective furnishing fabric that is solid enough to display wool in an interesting way. Here I have used the most basic stitches to 'draw' on fabric. Using this method you can reproduce children's drawings in embroidery or, indeed, any memorable scribbles that you wish to save for posterity.

Preparation

Enlarge the chart overleaf by 142 per cent and then, using the pencil, trace off the outline. Turn the tracing over and draw over the lines with a transfer pencil. Make a 4 cm (1½ in) hem at one short end of the calico to form the top of the bag and then transfer the image to the centre of the hemmed calico using a hot iron.

Stitching

Using half thickness of wool throughout, outline the cow in black (9800) back stitch. Continuing with the black, embroider the trellis pattern with single straight stitches secured at each cross section with a small horizontal straight stitch. Fill in the udder in the same way but using pink (8432). Outline the eye in black (9800) back stitch and fill in the pupil with two small rust (8236) straight stitches. Outline the nostrils in black (9800) back stitch and place a pink (8432) colonial knot in the centre of each. Work the chicken outline in back stitch using rust (8236) then add the comb in red (8202) satin stitch. Fill in the wing with criss-cross straight stitches in rust (8236). Embroider a beak with two angled, yellow (8154) straight stitches. Embroider 'Moo' in red (8202) back stitch. Finally, work the grass in lime (9152) straight stitch and the flower in red (8202) detached chain. Thread the bell onto the ribbon and stitch the ribbon ends into position below the cow's neck.

EMBROIDERY MATERIALS

STITCHES (SEE PAGES 110-113)

Back

Colonial knot

Detached chain

Satin

Straight

MATERIALS

Pencil

Tracing paper

Transfer pencil

1 piece of calico measuring 48 x 53 cm (19 x 21 in)

Medium-sized crewel needle

Small bell

2.5cm (1 in) of narrow ribbon

1 piece of ticking or heavy cotton measuring 48 x 53 cm (19 x 21 in)

Sewing thread to match calico

2 strips of matching fabric (for handles) each measuring 117 x 18 cm (46 x 7 in)

Pins

2 large buttons

YARNS

Anchor Tapisserie Wool

1 skein of each of the following colours:

Black 9800

Rust 8236

Red 8202

Lime 9152

Pink 8432

Yellow 8154

Split all wools in two and use as half thickness.

Making up

Stitch a 4 cm (1½ in) hem at one short end of the backing fabric (the ticking or heavy cotton) to form the back top. Press the back and front pieces under a damp cloth. Lay the calico and backing fabric right sides together and, allowing a 2.5 cm (1 in) seam allowance, hand sew or machine around the sides and bottom seams. Turn right sides out.

To make the handles, take one of the narrow strips of backing fabric and fold it in half lengthwise right sides together. Machine or hand sew the seam taking a 1 cm (½ in) allowance. Stitch across the bottom seam then turn the handle the right side out and press. Hand sew the remaining opening closed, tucking in the raw edges to neaten. Repeat for the second handle. Pin one handle end onto the right side front of the bag approximately 5 cm (2 in) in from the edge and 7.5 cm (3 in) down from the top. Stitch into place and add a decorative button. Repeat with the other end of the handle to the left front side of the bag. Attach the back handle as for the front, but don't add buttons.

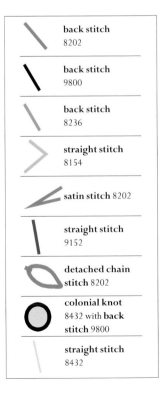

back stitch 8202
back stitch 9800
back stitch 8236
straight stitch 8154
satin stitch 8202
straight stitch 9152
detached chain stitch 8202
colonial knot 8432 with **back** stitch 9800
straight stitch 8432

Christmas Stocking

STITCHES (SEE PAGES 110-113)

Blanket

Colonial knot

Feather

Satin

Straight

MATERIALS

1 piece of red felt measuring
60 cm (24 in) square

Pencil

Tracing paper

Pins

1 piece of pure wool blanketing
measuring 12.5 x 50 cm (5 x 20 in)

Water-soluble pen

Medium-sized crewel needle

White sewing thread

30 cm (12 in) length of ribbon

YARNS

Anchor Tapisserie Wool

1 skein of each of the
following colours:

Red 8200

Yellow 8122

Dark brown 8410

Fawn 9404

2 skeins of Green 8972

DMC Fil Argent

1 reel of Silver Art 277

T his is a jolly Christmas stocking to hang above the fireplace and your chance to practise using feather stitch. This pretty stitch is used widely in smocking and is very simple to work. It is also a useful stitch for joining patches of crazy patchwork. I have joined the back and front of the stocking with decorative blanket stitch. If you intend to substantially stuff the stocking with goodies, I suggest you machine seam around the edge and then work the blanket stitch over the top.

PREPARATION

Fold the felt in half. Trace off the outline of the stocking following the dotted line on the chart overleaf. Cut two stocking shapes from the red felt. Trace off the star shapes and cut them out of paper. Pin them into position on the front of the stocking and draw around them lightly in pencil. Do not use water-soluble pen as it is difficult to remove from felt.

Fold the blanketing strip in half widthwise. Trace off the Christmas trees and pin to the centre of the folded blanketing. Mark guidelines through the tracing with the water-soluble pen. Open the blanketing out before you start sewing.

STITCHING

Work the tree barrels in fawn (9404) satin stitch following the direction of the stitches as indicated on the chart. Add two horizontal dark brown (8410) straight stitches to each barrel, then work the tree trunks in dark brown vertical straight stitches. Starting at the top of the tree and using green (8972), work feather stitch on either side of the tree trunks. Add the central band of feather stitches down each trunk. Work the red (8200) and yellow (8122) tree decorations in colonial knots, wrapping the wool twice around the needle for each knot. Add the straight stitch silver (277) star and the straight stitch tinsel as positioned on the chart.

Return to the red felt stocking front and stitch the stars using one silver straight stitch for each angle.

MAKING UP

Lay the two stocking pieces face down on the worksurface and next to each other as in the diagram, left. Take the embroidered blanketing and lay it on top of the stocking pieces embroidery face down and raw edges aligned. The stockings and blanket need to be arranged so that the space between the stockings is twice the excess fabric at each side of the stocking, less a 1 cm (½ in) seam allowance at each raw end. Pin and then hand or machine stitch the two layers together.

Turn the blanketing up and then place the stocking pieces together with wrong sides facing and join them together with a continuous row of green (8972) blanket stitches. Stitch together the short ends of the blanketing taking a 1 cm (½ in) seam allowance and fold the stocking top down over the red felt. Tuck the bottom edge of the blanketing underneath itself and catch all around the stocking. Fold the ribbon in half to form a 15 cm (6 in) loop and stitch it to the top right-hand corner of the stocking.

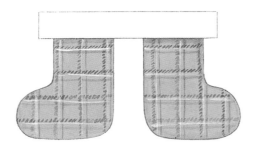

All stitches are worked in Anchor Tapisserie Wool unless otherwise described			
	straight stitch DMC Fil Argent Silver Art 277		straight stitch DMC Fil Argent Silver Art 277
	feather stitch 8972		satin stitch 9404
	straight stitch DMC Fil Argent Silver Art 277		straight stitch 8410
	colonial knot 8200		straight stitch 8410
	colonial knot 8122		

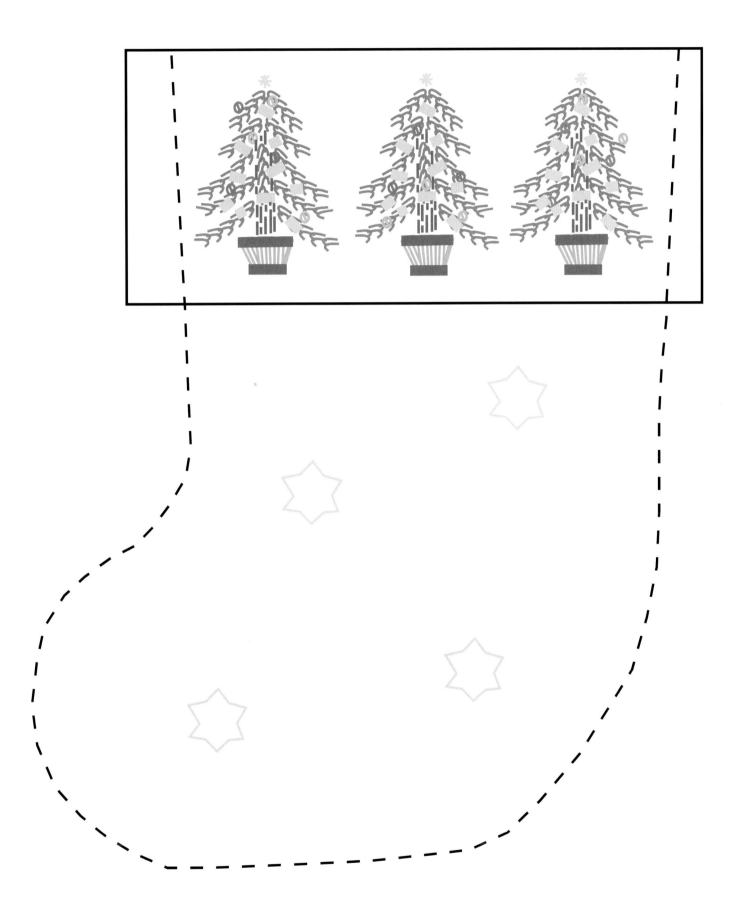

Perfect Pastels

This is undoubtedly my favourite section as I love the nostalgic feel of soft, pastel stitchery on a woolly background. The following items would all make perfect gifts for someone worth the trouble and can be stitched relatively quickly. If the making-up process worries you, consider working the designs on ready-made items such as towelling (terry cloth) slippers and cosmetic bags that can be bought cheaply. Substitute wools for ribbons or vice versa and add or subtract flowers to create your dream garden on fabric. Look in the Flower and Motif Libraries (pages 114-115 and 116-122) for extra ideas, but keep your colours pale and pretty.

Embroidered Rabbit

The floppy ears of this rabbit are a perfect invitation for wool embroidery. To make him extra glamorous, I have highlighted the flowers with some perlé cotton. You can put as much or as little embroidery as required into this project and you can make the rabbit himself from either wool blanketing or calico. My rabbit's whiskers were actually donated by a friendly local horse who allowed me to snip some hairs from his tail. If you don't happen to have a horse in your garage, nylon thread would work equally well.

PREPARATION

Enlarge the pattern pieces from the templates on pages 124-125 by 200 per cent and transfer these outlines to brown paper. Cut out the shapes and pin them to the fabric. Cut out the fabric pieces. Also enlarge the charts overleaf by 167 per cent.

STITCHING

Outer ears Trace the embroidery onto tissue paper and pin into position on the fabric. Mark the centres of the flowers through the paper with the water-soluble pen and work the flowers from the chart noting from the yarns list, left, which wools should be separated into half thicknesses and which should be used direct from the skein. Begin with the central hollyhocks (see page 114) and work out from here. The embroidery on the right ear is a mirror image of that on the left.

Inner ears With the water-soluble pen draw a large teardrop shape at the centre of the inner ear pieces and work a trellis of crossed straight stitches in coral (8306). Secure each cross with a small crossed straight stitch in medium pink (8364).

Hind legs Transfer the pattern onto the outer haunches as for the ears and work both haunches as indicated on the charts using perlé cotton and wool as required.

Body On each front paw work the daisies, forget-me-nots and leaf spray as indicated on the charts.

Tail Stitch the embroidery onto the tail piece as indicated on the chart.

Head Work the yellow daisies, pink forget-me-nots and leaf sprays only.

MAKING UP

NB There is a 1 cm (½ in) seam allowance included on all pieces of fabric.

Ears Lay the inner ear on top of the outer ear with wrong sides together. Turn in the edges of the outer ear so they meet the edges of the inner ear and hand stitch into place turning under the seam allowances. Join the horizontal seam above the inner ear.

Head Place the two halves of the head right sides together. Machine or hand stitch from points A to B. Turn right sides out and pin the narrow ends of the ears to the inside of the head pieces in the positions indicated by the dotted line. Place the point of the upper head marked C adjacent to A and pin this piece into position sandwiching the ears into place. Hand sew in place from A to D.

Using medium pink (8364), embroider a triangle of satin stitch over the seam at the point marked X to form the nose and then three straight stitches over the seam to form the mouth. Complete the embroidery on the rabbit's forehead, working over the seams. Stuff the head firmly and then make the eyes. Using plum (8330), embroider one eye in satin stitch in the position marked '0'. When this is

All stitches are worked in Anchor Tapisserie Wool unless otherwise described			
lazy daisy stitch 8814 with **colonial knot** 8040	**lazy daisy stitch** 8364	**colonial knot** Anchor Perlé 300	**colonial knot** 8932/8112
colonial knot 8040	**colonial knot** 8364/Anchor Perlé 300	**colonial knot** Anchor Perlé 336	**buttonhole wheel** 8368
colonial knot 8602	**detached chain stitch** Anchor Perlé 278	**buttonhole stitch** Anchor Perlé 278	**buttonhole wheel** 8366
colonial knot 8342	**lazy daisy stitch** 8012 with **colonial knot** 8112	**straight stitch** Anchor Perlé 336 with **colonial knot** Anchor Perlé 295	**buttonhole wheel** 8364
straight stitch 8364	**satin stitch** 8330	**colonial knot** 8342/8040	**buttonhole stitch** 8040
feather stitch Anchor Stranded Cotton 265	**lazy daisy stitch** Anchor Perlé 300 with **colonial knot** Anchor Perlé 295	**tête de boeuf** 8932	**straight stitch** 8306/8364
lazy daisy stitch 8040 with **colonial knot** 8364	**colonial knot** 8814/8112	**fly stitch** Anchor Stranded Cotton 265	

complete, take the thread through to the other side of the stuffed head, pull taut and secure it with a knot so that the eyes are slightly indented. Working over the knot, complete the second eye in satin stitch. To make the whiskers, thread lengths of nylon thread (horse hair) through the head and knot on each side to secure.

BODY

Take the three body pieces and, with right sides together, stitch the under body to the two body pieces. Join the back seam of the two body pieces leaving an opening from E to F to insert the head. Turn the body the right side out and stuff it taking care to push the stuffing down into the paws. Hand stitch the head to the body adding stuffing as necessary to keep it firm.

Hind legs Take the inner and outer haunch of each leg and place these with the right sides together. Machine or hand stitch together leaving a small opening. Turn right side out, stuff lightly and secure the opening. Pin and hand sew the haunches to the body from points G to H in the direction of the arrows.

Tail Make a row of running stitches around the seam allowance on the tail piece. Gather slightly and insert the stuffing. Pin the tail into position at the back of the rabbit and hand stitch into place. Tie a ribbon around the neck and make a bow.

Outer ear

Tail   Inner ear

Nose

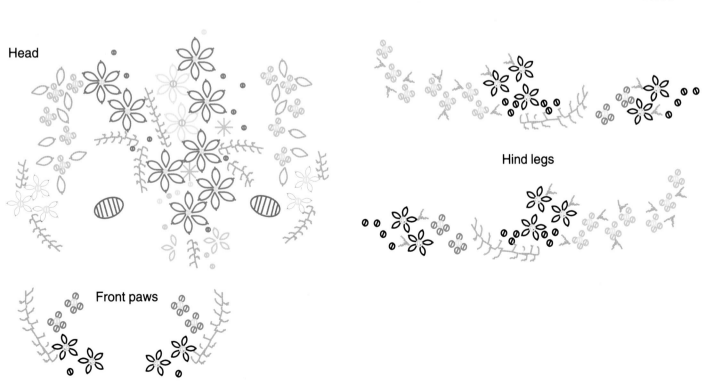

Head

Hind legs

Front paws

Floral Blanket

This pretty rug is made up of a series of simple repeated and mirrored patterns. The edging is bound in satin ribbon, but the edges of the blanket could be frayed or trimmed with lace. The simplest way to transfer this pattern is with tissue paper. Trace off the necessary number of repeats and baste the panels to the blanket. As the image is intricate, it is advisable to work the actual stitches with the tissue in place and to tear it away on completion.

PREPARATION

Trace off two images and two mirror images of Motif A overleaf, and position these in opposite corners placing them approximately 14 cm (5½ in) in from the edges of the fabric. Trace off four images and four mirror images of Motif B, also overleaf: two images (one mirrored) are used for each side. Place these between the main motifs to form a border. Baste the tracings into position.

STITCHING

Following the colour key accompanying the charts, begin with the bullion roses in Motif A. Start at the centres and shade the stitches from dark to light. Work the sprays of fly stitches around the roses keeping the stitches very close together. Then work the five open sprays of fly stitches around the outside of the design.

The forget-me-nots are worked in detached chain stitch in shades of blue with four petals to each flower. Following the chart for colour choice and position, work these in a circle from a central point and add a colonial knot to the centre of each. Forget-me-not leaves are single fly stitches. The white daisies are worked in the same way as the forget-me-nots but with five detached chain stitches for each flower head. The feathery flowers by the roses are worked in extended colonial knots. The group of three or four stitches join at the centre and are covered with a colonial knot. The pansies are made in satin stitch with the two shades of lavender.

MAKING UP

When all the embroidery is finished, cut the ribbon equally into four and fold each in half lengthwise. Pin along the edges of the blanket so that the ribbon covers the front and back. Fold in the corners to form mitres and hand stitch to neaten. Machine or hand stitch the edges of the ribbon onto the blanket.

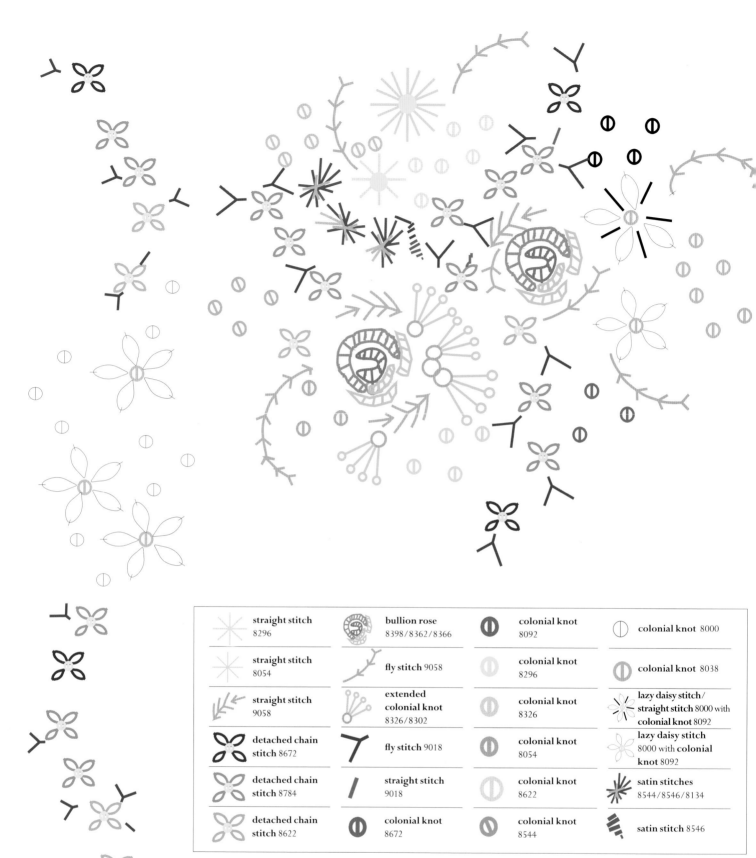

	straight stitch 8296		bullion rose 8398/8362/8366		colonial knot 8092		colonial knot 8000
	straight stitch 8054		fly stitch 9058		colonial knot 8296		colonial knot 8038
	straight stitch 9058		extended colonial knot 8326/8302		colonial knot 8326		lazy daisy stitch / straight stitch 8000 with colonial knot 8092
	detached chain stitch 8672		fly stitch 9018		colonial knot 8054		lazy daisy stitch 8000 with colonial knot 8092
	detached chain stitch 8784		straight stitch 9018		colonial knot 8622		satin stitches 8544/8546/8134
	detached chain stitch 8622		colonial knot 8672		colonial knot 8544		satin stitch 8546

Daisy Bag

EMBROIDERY MATERIALS

STITCHES (SEE PAGES 110-113)

Colonial knot

Detached chain

Feather

Fly

Lazy daisy

MATERIALS

1 piece of pure wool blanketing measuring 63 x 31 cm (25 x 12 in)

Water-soluble pen

Medium-sized crewel needle

31 cm (12 in) length of 15 mm (⅝ in)-wide pink bias binding

Pink sewing thread

Pins

1 m (1 yd) of narrow satin-faced ribbon

2 beads

YARNS

Anchor Tapisserie Wool

1 skein of each of the following colours:

Pink 8363

Pale pink 8342

Yellow 8112

Green 9055

White 8000

Split all wools in two and use half thickness

T his pretty little bag is especially suitable for beginners. It is easy to make and only uses three basic stitches. The right side of the heart is a mirror image of the left side, so while you could transfer the pattern accurately, it is just as easy to work the first half freestyle and then copy the positions of your flowers onto the second half. Once completed, it is the perfect bag in which to keep hair accessories, jewellery or baby items.

INSTRUCTIONS

Enlarge the heart template overleaf by 125 per cent, transfer the outline onto a piece of paper and cut out the shape. Fold the fabric in half widthways and position the heart cut-out centrally, with the bottom point 5 cm (2 in) up from the fold in the fabric. Draw around the heart with the water-soluble pen and remove the paper. Cut the wool into manageable lengths and carefully split each length in two so that you are working with half the normal thickness. Embroider the large, pink (8363) daisies in lazy daisy stitch, positioning each stitch centrally on your drawn line in the places indicated on the chart. Work the pale pink (8342) daisies next and then the white (8000) daisies. Fill in each flower centre with a yellow (8112) colonial knot. Next, work the leaves. Each one consists of one fly stitch (9055). To complete the curve at the top of each heart, work the spray of buds with fly stitch (9055) leaves filled in-between by two small pink (8363) straight stitches. Fill in any jagged or unevenly spaced areas with white (8000) colonial knots. Leaving 10 cm (4 in) of fabric free at the top of your design, surround the heart with random daisies and colonial knots in the two shades of pink.

FINISHING

Fold the fabric in half with right sides facing and stitch up the side seams 2.5 cm (1 in) from each edge. Turn right sides out. Fold and pin the bias binding over the top edge and hand sew this carefully into place with the pink sewing thread. Cut the length of ribbon in two, thread one end into the crewel needle and make a straight row of running stitches across the front of the bag, 6.5 cm (2½ in) from the top opening. Repeat with the remaining ribbon on the back of the bag. On each side, thread the two ends of ribbon through a bead and knot the ribbon above and below the bead to secure.

fly stitch 9055 with **straight stitch** 8363	colonial knot 8112
lazy daisy stitch with **colonial knot** 8363	fly stitch 9055
lazy daisy stitch 8000 with **colonial knot** 8112	colonial knot 8342
lazy daisy stitch with **colonial knot** 8042	colonial knot 8000
detached chain 9055	feather stitch 9055

Floral Cushion

In addition to wool, this design includes touches of silk ribbon embroidery, buttons and beads. If you don't want to make up the cushion from scratch, use the motifs on a ready-made plain cushion cover. The ribbon roses are created from straight stitches worked in a circle and you will find step-by-step instructions for making these in the Flower Library on page 115.

PREPARATION

Using the water-soluble pen and a compass, draw a 20 cm (8 in) diameter circle on the centre of the blanketing. Lay the tracing paper over the chart, overleaf, and mark the positions of the ribbon roses – aqua, pink with rose centre, mid-blue and peach. Place this over the fabric within the circle and push the pen through the paper to indicate the positions of these flowers.

STITCHING

Stitch the pink ribbon roses first referring to the straight stitch rose instructions in the Flower Library on page 115. Then work the large and small peach ribbon roses and the colonial knot rosebuds following the chart for choice of colours. Once the pink and peach roses are complete, work out from here following the chart by eye and completing the remaining flowers in the circular image using both ribbons and wool. When you work the silk ribbon leaves note that the sprays of fly stitch are worked in 2 mm (⅛ in) ribbon while the larger leaves are in 4 mm (⅙ in) ribbon. Instructions for the yellow and rose honeysuckles stitched with 2 mm (⅛ in) ribbon can be found in the Flower Library on page 115. Sew the pearls into position with sewing thread and attach the buttons with pale lilac (8582) wool and a fine needle, working two colonial knots (one above each hole) to secure.

The corner motifs are worked as mirror images. Position the aqua ribbon roses 6.5 cm (2½ in) in from the top (or bottom) and side edges. Work the light and

dark green ribbon leaf sprays in fly stitch and then add the lilac forget-me-nots, pink daisies and pink colonial knots as indicated in the chart. Stitch on the buttons as described for the circular motif and then work the three small peach colonial knot buds. Finally, add the three aqua ribbon detached chain stitches.

MAKING UP

Lay the piping cord in the centre of the wrong side of the fabric strip. Fold the fabric in half and machine or hand stitch a continuous row of back stitches through both thicknesses of fabric as close as possible to the piping cord. Turn under a 6 mm (¼ in) selvedge at all sides of both pieces of backing fabrics. Machine seam or hand stitch into place. Do the same with the raw edges of the blanketing. Lay the embroidery face down and pin one of the backing pieces to the top half, sandwiching the piping in between. Then, keeping the piping in place, hand stitch the sides and top closed. Repeat for the bottom half of the backing but allow this to overlap the top half so leaving a centre back opening to insert the cushion pad.

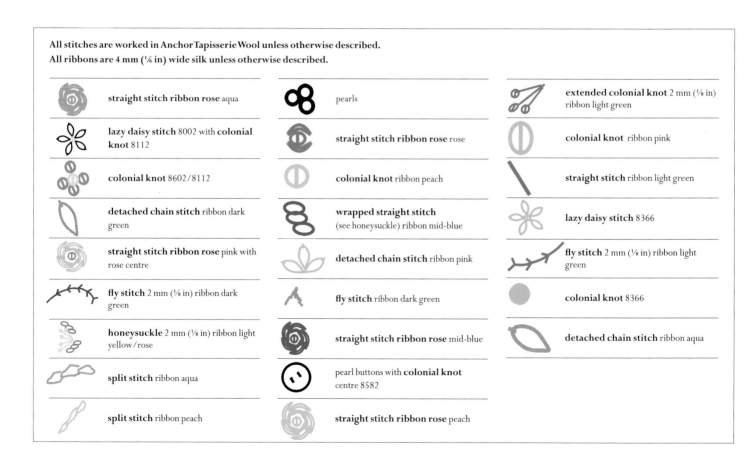

All stitches are worked in Anchor Tapisserie Wool unless otherwise described.
All ribbons are 4 mm (⅙ in) wide silk unless otherwise described.

straight stitch ribbon rose aqua	pearls	extended colonial knot 2 mm (⅛ in) ribbon light green
lazy daisy stitch 8002 with colonial knot 8112	straight stitch ribbon rose rose	colonial knot ribbon pink
colonial knot 8602/8112	colonial knot ribbon peach	straight stitch ribbon light green
detached chain stitch ribbon dark green	wrapped straight stitch (see honeysuckle) ribbon mid-blue	lazy daisy stitch 8366
straight stitch ribbon rose pink with rose centre	detached chain stitch ribbon pink	fly stitch 2 mm (⅛ in) ribbon light green
fly stitch 2 mm (⅛ in) ribbon dark green	fly stitch ribbon dark green	colonial knot 8366
honeysuckle 2 mm (⅛ in) ribbon light yellow/rose	straight stitch ribbon rose mid-blue	detached chain stitch ribbon aqua
split stitch ribbon aqua	pearl buttons with colonial knot centre 8582	
split stitch ribbon peach	straight stitch ribbon rose peach	

EMBROIDERY MATERIALS

∞

STITCHES (SEE PAGES 110-113)

Bullion

Colonial knot

Lazy daisy

Satin

Stem

Straight

MATERIALS

Pencil

Tissue paper

Pins

1 piece of pure wool blanketing
measuring 20 x 12.5 cm (8 x 5 in)

Contrasting thread

Medium-sized crewel needle

Water-soluble pen

1 piece of furnishing fabric measuring
50 x 25 cm (20 x 10 in)

1 piece of medium-weight iron-on
interfacing measuring 20 x 12.5 cm
(8 x 5 in)

50 cm (20 in) of piping cord

Sewing thread to match fabric

2 pieces of pinked felt each measuring
5 x 10 cm (2 x 4 in) (optional)

50 cm (20 in) narrow ribbon (optional)

YARNS

DMC Medici Crewel Wool

1 skein of each of the following colours:

Light brown 8501

Medium brown 8611

Pastel pink 8818

Light pink 8225

Medium pink 8224

Pale green 8369

Light green 8871

Pale yellow 8328

Very pale gold 8327

Medium gold 8325

Charcoal 3021

Harvest Mouse Sewing Case

∞

This little pouch was designed to be useful. It is the right size to keep your specs in or, alternatively, you can use it as a sewing tidy, adding some felt squares for your needles. The little mouse is the work of Jillian Taylor who has used a similar mouse on a baby's cot (crib) set. You could use the mouse as an alternative to the teddy bears which appear on the blanket on pages 98-101.

PREPARATION

Trace the chart overleaf onto tissue paper and pin it to the centre of the blanketing. Baste over the outline in contrasting thread and remove the paper.

STITCHING

Starting with the fur, work in straight stitches following the colours and directions given on the chart. Fill in the light patch on the mouse's ear with pastel pink (8818) straight stitches and overlay a few medium brown (8611) stitches to soften the line. Also work the tip of the mouse's nose in medium brown satin stitch. Fill in the mouse's jacket in light pink straight stitch (8225) following the direction of the stitches shown on the chart. Work the stems of the wheat in sloping satin stitch following the chart for the appropriate colours.

Outline the mouse and jacket in stem stitch using medium brown (8611) and medium pink (8224). Note the extra row of light brown (8501) stem stitch outlining the pale pink ear. Work a medium pink (8224) colonial knot for the button on the collar. Following the chart for the colours, work the ears of wheat in bullion stitch with the wool wrapped around the needle six or seven times. Criss-cross the straight stitches by the wheat ears to give a natural effect. Work the grass in straight stitch in light and pale green (8871 and 8369) and the lazy daisies in the light and medium pink (8225 and 8224) finishing each centre with a pale green (8369) colonial knot. Add the whiskers and eyes in straight stitch using charcoal (3021).

MAKING UP

Lay the finished piece face down on a soft towel and press over a damp cloth. Turn
to the right side. With the water-soluble pen, draw a rectangle measuring
20 x 12 cm (8 x 4¾ in) around the design, taking care to centre it. Trim off excess
blanketing leaving a 1.5 cm (⅝ in) selvedge. Cut a piece of backing fabric to match
the wool blanketing and iron the interfacing onto the wrong side.

satin stitch 8325		straight stitch 8369	
satin stitch 8369		straight stitch 8818	
bullion stitch 8369		straight stitch 3021	
bullion stitch 8325		straight stitch 3021	
bullion stitch 8328		straight stitch 8501	
stem stitch 8611		straight stitch 8225	
stem stitch 8224		stem stitch 8501	
lazy daisy stitch 8225 with colonial knot 8369		straight stitch 8325	
lazy daisy stitch 8224 with colonial knot 8369		straight stitch 8327	
		colonial knot 8224	
straight stitch 8871			

Cover the piping by cutting a 46 cm (18 in) strip of furnishing fabric 4 cm (1½ in) wide and double this over the piping cord with wrong sides together. Machine or hand seam along the doubled strip so securing the cord neatly in the fold.

With the embroidery facing you and starting at the left end with raw edges aligned, carefully baste the piping cord to the wool blanketing around two long sides of the rectangle and one short end, leaving the other end un-piped. Snip the covering of the piping at the two corners so that it carries around the corners with ease. Clip away the ends of the piping cord inside the fabric tube to give a neat finish. Place the fabric tube and embroidery with right sides together and stitch around the three piped sides keeping close to the piping. Trim off waste fabric leaving a 6 mm (¼ in) allowance. Zigzag the raw edges and turn right sides out.

For the lining, cut a piece of fabric 12 x 36 cm (4¾ x 14 in) and fold lengthwise in half. Stitch along both long edges taking a 1 cm (½ in) seam allowance. Insert the lining in the case without turning. Turn in the top edges and slip stitch together. Catch the two felt squares together in the top left-hand corner. Tie a small bow in the end of the ribbon and stitch the bow over the corner of the felt. Stitch the other end of the ribbon to the inside top edge of the pouch.

Floral Mules

≈

This design could be used to decorate shop-bought towelling mules (terry cloth slippers) or you could follow the simple instructions given overleaf and make your own. I have stitched a shop-bought leather inner sole onto the base of each sole. As an alternative, you could cut and glue any non-slip material into place, or baste coils of heavyweight string onto the base to stop yourself sliding around.

PREPARATION

These mules are designed to fit foot size 5-6 and to give you the basic shape of the required pieces. The size can be simply adjusted by lengthening or shortening the pieces. Pin the pieces together before lining and adjust to the size of your foot.

Enlarge the templates given overleaf by 200 per cent and cut four sole pieces and two top pieces from the blanketing fabric. Use the top side of the fabric for the right foot and turn it over, so the shaping is reversed, for the left foot. Cut two top pieces from lining fabric.

Trace off the design twice onto tissue paper and pin one tracing right side up on the right foot and the other, wrong side up on the left foot. With the tip of the water-soluble pen, push through the paper and mark the centres of the main flowers. Cut the wool into manageable lengths and carefully split each length in two so you are working with half the normal thickness.

STITCHING

Following the chart, begin by working the three bullion roses. Stitch the three leaves around the roses using diagonal straight stitches placed close together and graduating at the edges to form the leaf shape. Next work the spray of buttonhole wheels in yellow (8036) ending with a quarter wheel in green (9076). Stitch the forget-me-nots in dark lilac (8590) working four colonial knots with a yellow

---

EMBROIDERY MATERIALS
≈

STITCHES (SEE PAGES 110-113)

Blanket

Colonial knot

Detached chain

Fly

Lazy daisy

Straight

plus, from the Flower Library
(see pages 114-115):

Bullion rose

MATERIALS

Pencil

Tracing paper

1 piece of pure wool blanketing
measuring 30 x 91.5 cm (12 x 36 in)

1 piece of cotton lining fabric
measuring 46 x 18 cm (18 x 7 in)

Water-soluble pen

Medium-sized crewel needle

1 pair of leather soles

Sewing thread to match lining fabric

Sewing thread to match slipper top

YARNS

Anchor Tapisserie Wool
1 skein of each of the following colours:

Dark pink 8414

Pale pink 8362

Light lilac 8588

Dark lilac 8590

Green 9076

Yellow 8036

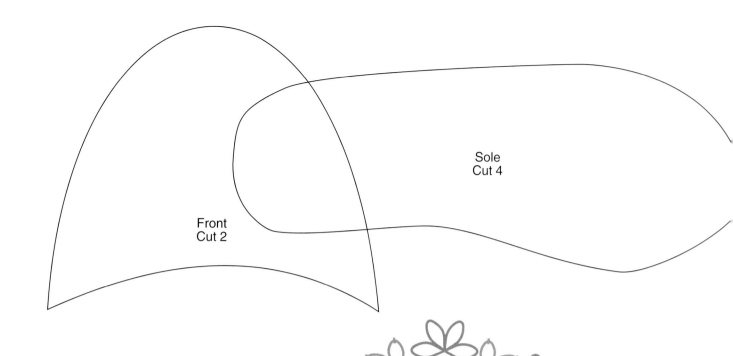

Front
Cut 2

Sole
Cut 4

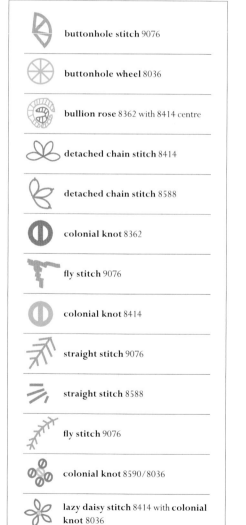

	buttonhole stitch 9076
	buttonhole wheel 8036
	bullion rose 8362 with 8414 centre
	detached chain stitch 8414
	detached chain stitch 8588
	colonial knot 8362
	fly stitch 9076
	colonial knot 8414
	straight stitch 9076
	straight stitch 8588
	fly stitch 9076
	colonial knot 8590/8036
	lazy daisy stitch 8414 with colonial knot 8036

(8036) knot in the centre. Fill in the dark pink (8414) detached chain flowers above the buttonhole wheels and at both bottom corners. Fill in the light lilac (8588) and pale pink (8362) lazy daisy stitches and the surrounding dark and pale pink individual colonial knots. Finally, work the fly stitch leaves in green (9076) beside the daisies.

MAKING UP

Take one blanket sole for each foot and hand or machine stitch – or glue – the leather soles to the undersides. Place these before you, leather soles down and lay the second pair of blanket soles on top of the first pair. Oversew these together around the outer edge, tucking in the raw edges of fabric as you go along.

Lay the embroidered uppers face down on the cut-out lining pieces. Seam three-quarters of the way around the outside edge, turn right sides out and complete the seam. Pin the lined uppers to the soles and join by oversewing firmly around the edges with a double thickness of sewing thread.

Classic Checks

Checked fabrics are fashionable and practical and provide the perfect backdrop for your embroidery. For a real country feel, I have included some of my farmyard friends who are guaranteed to brighten up even the dreariest morning. If you have a checked blanket or curtain that is a little worse for wear, cut it up and make yourself some embroidered items from the good bits. If you run out of fabric you can always back cushion panels with a plain wool or calico. If you have any leftover scraps, what about making egg cosies or baby mittens or bootees? Look at the next section in this book for some motif ideas.

EMBROIDERY MATERIALS

STITCHES (SEE PAGES 110-113)

Buttonhole

Chain

Cretan

Satin

Split

Straight

MATERIALS

Pencil

Tissue paper

Pins

1 piece of pure wool blanketing
measuring 92 x 60 cm (36 x 24 in)

Sewing thread

Medium-sized crewel needle

25 g (1 oz) ball of double
knitting wool

YARNS

Anchor Tapisserie Wool

1 skein of each of the
following colours:

Orange 8166

Bright red 8198

Dark red 8202

Yellow 8018

Butter 8012

Dark green 8972

Lime green 9156

Fruit Salad Picnic Rug

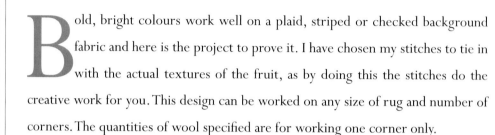

Bold, bright colours work well on a plaid, striped or checked background fabric and here is the project to prove it. I have chosen my stitches to tie in with the actual textures of the fruit, as by doing this the stitches do the creative work for you. This design can be worked on any size of rug and number of corners. The quantities of wool specified are for working one corner only.

PREPARATION

Trace off the outline of the chart overleaf onto tissue paper. Pin the paper to the corner of the rug and baste over the basic shapes. Remove the tissue paper.

STITCHING

Begin with the oranges, working a split-stitch outline in butter (8012) and then divide the segments with rows of split stitch. Around each piece of orange work a row of split stitches along the outer edge in orange (8166) and then fill in the segments in orange chain stitch. Do the same with the limes using the appropriate colours on the chart.

Next work the banana slices: these consist of full and half buttonhole wheels using blanket stitch. Once the wheels are stitched, thread some yellow (8018) wool and, starting at the outer edge, weave it in and out of the spokes without going through the fabric. Work in a spiral, fastening off at the centre of the circle. Outline the strawberries with dark or bright red (8202 and 8198) split stitch as indicated on the chart. Fill in the fruits with random straight stitch in the two shades of red, covering the split-stitch outline as you go. Add the pips randomly in straight stitch following the colours on the chart. Take care not to pull the threads through too tightly so they remain above the red stitches. Work the small leaves at the top of the strawberries in dark green (8972) satin stitch. Next work the orange leaves in dark green (8972) Cretan stitch.

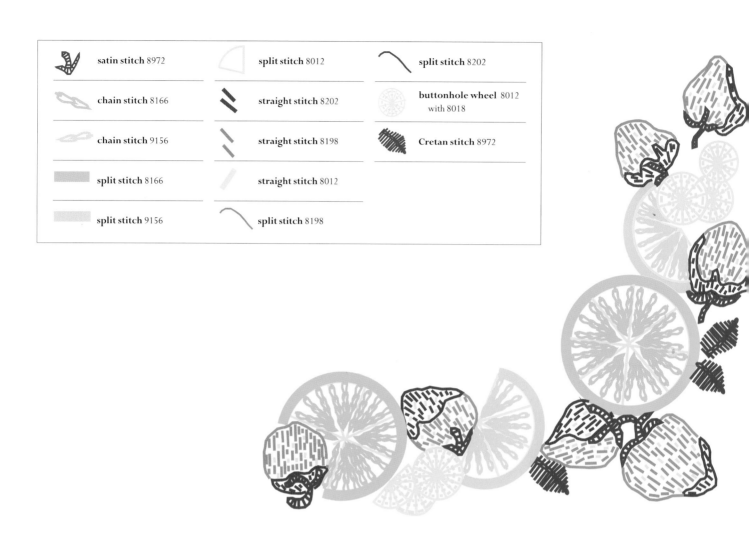

satin stitch 8972	split stitch 8012	split stitch 8202
chain stitch 8166	straight stitch 8202	buttonhole wheel 8012 with 8018
chain stitch 9156	straight stitch 8198	Cretan stitch 8972
split stitch 8166	straight stitch 8012	
split stitch 9156	split stitch 8198	

MAKING UP

When the design is complete, fold back the edges of the blanket to form a 1.5 cm (⅝ in) hem and baste into position. Stitch all around the edge with blanket stitch using the knitting wool.

Pig Cushion

As the proud owner of two Vietnamese Pot Bellied Pigs I had to include a pig of some kind in this book. Here we have a mama pig enjoying the delights of a summer flower garden. In addition to wool I have used silk ribbons and beads to ensure that a touch of class is evident. She is, after all, a most particular pig.

PREPARATION

Trace off the outline of the pig from the chart on page 60 onto tissue paper. Pin the tracing to the centre of the white blanketing and baste around the shape in contrasting sewing thread. Tear away the tracing paper.

STITCHING

Pig Using dark peach (8324) wool and split stitch, work around the outline of the pig. Change to peach (8252) and work the cheek in circles of chain stitch. Make a colonial knot in black (9800) for the eye. Now fill in the pig with straight stitch using the peach and dark peach wools and following the direction of the stitches indicated on the chart. When the pig is complete, work a row of split stitches in pink (8396) along the underbelly and add seven colonial knots for the nipples.

To make the tail, tightly twist a single thread of peach (8252) wool. Fold it in half and let the two lengths twist together. Thread the twisted length through the needle and take it down through the blanketing at the centre of the back end of the pig, leaving 4 cm (1½ in) of twisted wool free at the front of your work. Knot at the back of the work and cut off the raw ends of wool.

FLOWERS

Work the lazy daisies in dark rose (8400) wool. Then work the straight stitch flowers using the rose 4 mm (⅙ in) wide ribbon at the rear end of the pig.

EMBROIDERY MATERIALS

STITCHES (SEE PAGES 110-113)

Blanket	Lazy daisy
Chain	Split
Colonial	Straight
Fly	Tête de boeuf

plus, from the Flower Library
(see pages 114-115):

Straight stitch rose

MATERIALS

Pencil

Tissue paper

1 piece of wool blanketing in cream
measuring 21 cm (8¼ in) square

Contrasting sewing thread

Medium-sized crewel needle

12 round glass beads

1 piece of check wool blanketing
measuring 41 cm (16 in) square

Ecru sewing thread

2 pieces of calico each measuring
41 x 28 cm (16 x 11 in)

164 cm (64 in) piping cord

1 strip of calico measuring
164 x 10 cm (64 x 4 in)

YARNS

Anchor Tapisserie Wool

1 skein of each of the following colours:

Peach 8252	Dark rose 8400
Dark peach 8324	Black 9800
Pink 8396	Blue 8644

Use half thickness of all colours except Blue

4 mm (⅙ in) wide silk ribbons

1 m (1 yd) rose	50 cm (½ yd) gold
1 m (1 yd) lilac	50 cm (½ yd) pale gold
50 cm (½ yd) indigo	50 cm (½ yd) light green

2 mm (⅛ in) wide silk ribbons

50 cm (½ yd) indigo	1 m (1 yd) bright green
50 cm (½ yd) pale indigo	50 cm (½ yd) violet

Following the chart by eye, continue to complete the flowers at the back of the pig using the ribbon colours and widths specified on the chart. When working the column of ribbon straight stitch in rose near the head, leave a loop of ribbon free, i.e. after inserting the needle, do not pull the ribbon right through the fabric. This also applies to the indigo ribbon straight stitches in the front flowerbed.

Next stitch the beads into position on the front flowerbed, securing each with ribbon in the appropriate colours. Complete the flowers as indicated on the chart.

MAKING UP

Pin the embroidered piece to the centre of the checked background panel and baste into place. Stitch all around the outer edge of the embroidered blanketing with blanket stitch in blue (8644), so joining the white panel to the check background panel. Remove the basting.

Take the strip of calico and fold it in half lengthwise over the piping cord with wrong sides facing and machine or hand seam through both thickness of calico very close to the cord. Fold back and pin a 1 cm (½ in) hem around the front fabric and the edges of the two back panels.

Pin the back panels together, overlapping the hemmed inner edges so that you have a 39 cm (15 in) square. At the same time, sandwich the raw edges of the covered piping cord between the outer edges of the back and front cushion panels. Baste through all thicknesses of fabric, then machine or handsew into place.

All stitches are in Anchor Tapisserie Wool unless otherwise described.
All ribbons are 2 mm (1/8 in) wide unless otherwise described.

fly stitch ribbon bright green	bead with ribbon lilac/lilac	**split** and **straight stitches** 8324	**ribbon stitch** 4 mm (1/6 in) ribbon indigo
fly stitch ribbon violet	**straight stitch** rose 4 mm (1/6 in) ribbon lilac	**straight stitch** 8396	**buttonhole stitch** ribbon bright green
ribbon stitch 4 mm (1/6 in) ribbon rose	**colonial knot** ribbon pale indigo	**colonial knot** 8396	**ribbon stitch** 4 mm (1/6 in) ribbon pale gold
ribbon stitch 4 mm (1/6 in) ribbon rose	**colonial knot** 4 mm (1/6 in) ribbon lilac	**lazy daisy stitch** 8400	**chain stitch** 8252
bead with ribbon indigo/indigo	**colonial knot** ribbon violet	**colonial knot** ribbon indigo	**ribbon stitch** 4 mm (1/6 in) ribbon gold
bead with ribbon lilac/indigo	**colonial knot** 9800	**tête de boeuf** ribbon indigo	**straight stitch** ribbon bright green
bead with ribbon lilac/rose	**straight stitch** 8252	**ribbon stitch** 4 mm (1/6 in) ribbon light green	

Cat Pincushion

While my own eight cats would be horrified at the idea, this is a pincushion for cat haters. You can stick pins in him *ad infinitum.* Kinder folk might like to stuff the cushion with sweet smelling herbs and hang him on a coat hanger. You can also change the colours so that he matches your own favourite moggy. Either way, he is easy to stitch, mostly using straight stitches.

PREPARATION

Trace off the cat outline onto tissue paper and pin this to the centre of a square of blanketing. Baste around the shape of the cat with white sewing thread showing the shape of the head and the division for the tail.

STITCHING

NB Divide the wool lengths into half thicknesses.

Outline the top of the head in white (8002) back stitch. Then work the eyes in white (8002) detached chain stitch, the nose and inside ears in pink (8304) satin stitch and the mouth in black (9800) straight stitches. Work the paws in vertical white (8002) satin stitch. Fill in the lower head, upper head, body and tail with straight stitches working in the colours and directions indicated on the chart. Build up the dense coat by working stitches on top of each other. Add the whiskers using one strand of the stranded cotton.

MAKING UP

Make a row of green thread running stitches on the bottom edge of the lace. Gather until it fits neatly around the outer edge of the cushion. Pin the gathered edge to the front of the embroidered panel aligning raw edges (see diagram, right). Baste into place and remove the pins. Place the back piece right side down on top

of the lace-trimmed front piece and machine or hand sew through the three thicknesses of fabric all around the outer edge, leaving 5 cm (2 in) open for turning. Turn right side out and remove basting. Stuff with polyester wadding (batting) and hand sew the remaining opening closed. Neatly join the short ends of lace.

Pin the velvet piping into position over the side seam and catch into place with matching thread. Neatly join the short ends. Take the remaining piping and fold in two to form a loop. Stitch into place behind the lace at the top right-hand corner of the cushion.

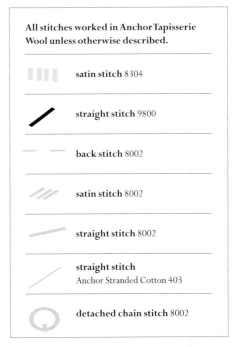

All stitches worked in Anchor Tapisserie Wool unless otherwise described.

satin stitch 8304	
straight stitch 9800	
back stitch 8002	
satin stitch 8002	
straight stitch 8002	
straight stitch Anchor Stranded Cotton 403	
detached chain stitch 8002	

Big Brown
Hen Cushion

This motherly hen is sitting pretty on a chair seat cushion made from a piece of tartan blanketing. She is modelled on Freda, the oldest of four Warren hens who range freely outside my studio window and obligingly supply me with breakfast. Like all the designs in this book, the project is worked from a small collection of stitches all of which can be learnt in a minute from the step-by-step guide on pages 110-113.

PREPARATION

Trace off the outline of the hen shape from the chart overleaf onto tissue paper. Pin this onto the centre of the front piece of the tartan fabric and baste over the tracing with white thread.

STITCHING

Remove the tissue paper and begin by working the head shape in straight stitch referring to the chart for guidance on the direction of the stitches and the colours. Next work the eye in back stitch and satin stitch and the beak in satin stitch, and then begin the wing feathers, working in sweeping rows of fly stitch. Start with the top curve of feathers, then work the bottom curve and finally fill in the area in between.

The trellis on the hen's tummy is stitched in zigzag rows of straight stitches. When you have completed these, join the diamond shapes at the bottom front curve of the tummy with individual straight stitches to create a rounded edge. Work the tail feathers in straight stitch and the feet in satin stitch.

Stitch the egg next in straight stitch and then work the ears of corn in detached chain and stem stitch. Once these are completed, add some randomly placed white lazy daisies with colonial knot centres among the corn and across the fabric.

MAKING UP

Cut out a paper template for the desired shape of your cushion and place this on the completed design, remembering to centre the design. Allowing a 2.5 cm (1 in) seam allowance around all four sides, cut out the front piece and repeat for the back. Round off the front corners with sharp scissors. Turn back and machine stitch a 2.5 cm (1 in) hem all around the embroidered piece of tartan. Hem the back piece to match.

Hand sew the piping into place around the right side of the front piece. Then slip stitch the back and front pieces together leaving a 25 cm (10 in) opening at the centre top for the padding. Insert the cushion pad. Finally, cut the ribbon into four equal lengths and insert the ends, back and front, flush to the top seam and stitch securely into place.

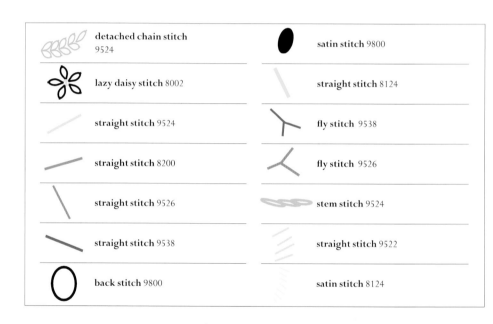

	detached chain stitch 9524		satin stitch 9800
	lazy daisy stitch 8002		straight stitch 8124
	straight stitch 9524		fly stitch 9538
	straight stitch 8200		fly stitch 9526
	straight stitch 9526		stem stitch 9524
	straight stitch 9538		straight stitch 9522
	back stitch 9800		satin stitch 8124

EMBROIDERY MATERIALS

❧

STITCHES (SEE PAGES 110-113)

Blanket   Lazy daisy

Colonial knot   Satin

Detached chain   Split

Fly   Straight

MATERIALS

Pencil

Tissue paper

2 pieces of pure wool blanketing
measuring 33 x 29 cm (13 x 11½ in)

2 pieces of cotton lining fabric the same
size as the blanketing

2 pieces of polyester wadding (batting)
the same size as the blanketing

Pins

Sewing thread (to match blanketing)

Medium-sized crewel needle

Water-soluble pen

13 flower-shaped buttons

YARNS

Anchor Tapisserie Wool

1 skein of each of the following colours:

Yellow 8094

Gold 8124

White 8002

Silver 8702

Magenta 8436

Butter 8038

Green 9156

Black 9800

Blue 8644

Anchor Perlé Cotton no. 5

1 skein of each of the following colours:

Fuchsia 77

Green 259

Anchor Stranded Cotton

1 skein of Green 265

Goose Tea Cosy

❧

Tea cosies are enjoying a revival and these two geese will add a touch of country living to any breakfast table. Worked here on blue-checked pure wool blanketing, you could work the same image on blue cotton gingham or perhaps on a fabric that matches your curtains or table linen.

PREPARATION

Trace off the tea cosy outline from the chart overleaf onto the tissue paper and use the template to cut two pieces from the blanketing and two from the lining fabric. Cut two shapes from the wadding (batting), 2.5 cm (1 in) smaller than the template. Trace off the outline of the geese and mark the centres of the flowers. Pin the tracing over the front tea cosy shape and baste around the geese outline. Mark through the tracing with a water-soluble pen to position the flowers.

STITCHING

In white (8002) split stitch, work around the outline of the geese. Fill in the body with straight stitches working two rows in silver (8702) to show the wing separation. Work the wings in blanket stitch using the white (8002). Stitch the beaks and feet in gold (8124) satin stitch. Stitch on the buttons using green cotton perlé (259) and still using the green, embroider the leaves and stems in detached chain and straight stitch. Work the lazy daisies in magenta (8436) wool and fuchsia cotton perlé (77) finishing the centres with colonial knots in butter (8038). The leaves are green wool (9156) straight stitches. The small white (8002) flowers are also worked in straight stitch spanning from a central point and finished with a colonial knot in butter (8038). The stems and leaves are straight stitch in green cotton perlé (259). Work the laburnum in yellow (8094) detached chain stitches. Add the fly-stitch leaf sprays using three strands of green stranded cotton (265).

Making up

Place the front and back lining pieces wrong sides together, and seam around the curved edge with a 1 cm (½ in) allowance. Place the embroidered front and back blanketing with right sides together and with the same allowance, seam around the curved edge and turn right sides out. Insert the two pieces of wadding (batting).

Insert the lining between the two pieces of wadding (batting). Open out the skein of blue (8644) wool and fold it into four lengths. Twist tightly then fold in half and let the two halves twist together to form a cord and secure the ends. Catch this cord around the curved edge of the tea cosy making a loop at the centre top. Pin the ends to the inside edge corners of the bottom of the tea cosy. Turn up raw edges of lining fabric and blanketing and slipstitch the seam around the bottom edge of the tea cosy to close, securing the twisted cord at the corners.

All stitches are worked in Anchor Tapisserie Wool unless otherwise described

detached chain stitch 8094	straight stitch 8002 with **colonial knot** 8038
fly stitch Anchor Stranded Cotton 265	**split** and **straight stitches** 8002
straight stitch Anchor Perlé 259	**straight stitch** 8702
button attached with Anchor Perlé 259	**blanket stitch** 8002
detached chain stitch Anchor Perlé 259	**straight stitch** 9156
lazy daisy stitch with **colonial knot** 8436/8038	**satin stitch** 8124
lazy daisy stitch with **colonial knot** Anchor Perlé 77/8038	**colonial knot** 9800

Monogrammed Cushion

These floral initials can be used in numerous ways. The images can be reduced or enlarged to fit any project and you can adapt the colours to fit in with your surroundings. On this design I have used two tones of pink on the lettering, shading the finest curve with a light tone. By varying the shades you can give the initials depth or make them appear three-dimensional; experiment by copying and filling in the outlines with crayon to see the different effects you can achieve.

PREPARATION

Trace off the letter of your choice from the charts on pages 75-79 onto tissue paper. Place the tracing centrally on the blanketing square and baste around the outline in white thread.

STITCHING

Embroider over the outline with slanting satin stitch using the perlé threads as shown on the chart and shading the colours from light to dark. Mark the position for each spider's web rose using the white chalk pencil. Work the spider's web roses as indicated on the chart. Work the rosebuds also in ribbons using either two or three small straight stitches crossing over each other at the base. Stitch the leaves in ribbon, crossing over the letter in places. Work the fly stitch leaves as indicated on the chart.

MAKING UP

Take the finished embroidery and lay a 25 cm (10 in) square with the tartan ribbon around the outside of the design, mitring each corner. Attach this to the fabric with the fusible webbing, following the manufacturer's instructions. Lay two loops of ribbon over the bottom right-hand corner of the square to form a mock bow.

Stitch this into place. Cut a 23 cm (9 in) length of velvet piping, fold it in half and twist the two halves together. With the cut ends tucked underneath, form a circle from the twisted piping so that it resembles a rosette. Stitch firmly into place.

Take a piece of tartan blanketing and make a 2.5 cm (1 in) hem along one long edge. Repeat for the other piece. Each piece should now measure 41 x 25.5 cm (16 x 10 in). Place the front right side up and lay the two back pieces face down on top, raw edges aligned, so that the hems overlap in the centre. Pin into position around the outside edge and then, allowing for a 2.5 cm (1 in) hem around the outside edge, hand sew or machine stitch into place. Turn the right way out. Using the dark green sewing thread, catch the piping all around the outer edge and sew the cut ends together. Insert the cushion pad.

spider's web rose dark rose	**straight stitch** green/dark rose	**straight stitch** green/old gold
spider's web rose old gold	**straight stitch** green/light rose	**fly stitch** green
spider's web rose yellow	**straight stitch** green/yellow	
spider's web rose light rose	**straight stitch** green	

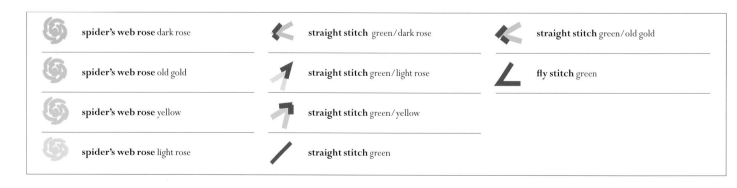

spider's web rose dark rose	**straight stitch** green/dark rose	**straight stitch** green/old gold
spider's web rose old gold	**straight stitch** green/light rose	**fly stitch** green
spider's web rose yellow	**straight stitch** green/yellow	
spider's web rose light rose	**straight stitch** green	

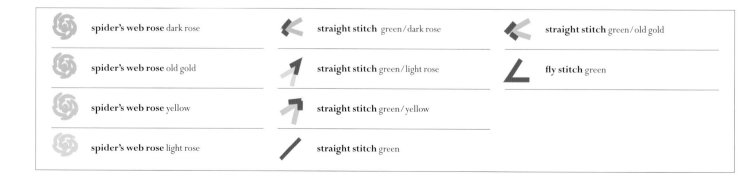

🌹	**spider's web rose** dark rose	◀	**straight stitch** green/dark rose	◀	**straight stitch** green/old gold
🌹	**spider's web rose** old gold	↗	**straight stitch** green/light rose	∟	**fly stitch** green
🌹	**spider's web rose** yellow	↗	**straight stitch** green/yellow		
🌹	**spider's web rose** light rose	╱	**straight stitch** green		

Sugar and Spice

The wonderful thing about projects for babies is their size. With small areas to cover, you can stitch really pretty items in just an evening or so. The projects in this section show you how to give plain clothes a very special look and also how to create really pretty gift items from left-over scraps of blanketing. Because the items are so small, I prefer to split the wool and use a half-thickness so that my embroidery looks delicate on the fabric. When filling in outlines for the chicks (pages 95-97) and teddy bears (pages 98-100), be careful to keep the stitching light and don't pull too tightly or you might distort the blanketing.

Bluebell Smock

This calico smock is designed to fit a 2 to 3-year-old and, following tradition, the back is left open and held together with ties. If you don't want to try your hand at dressmaking, use the woolwork design on a plain shop-bought pinafore dress or smock. Calico should be pre-washed before you cut your pattern as it tends to shrink.

PREPARATION

Enlarge the pattern pieces on page 123 by 400 per cent and transfer the outlines onto the calico and contrasting cotton fabric as indicated on the pattern. With right sides together, pin and sew the front bodice to the front skirt at the waistline. At the waistline, place vertical pins 7 cm (2¾ in) in from each edge to mark the position of the first and last flower and space ten pins evenly in between. Place two pins on the pocket to mark the bees' position. Finally, place a row of three pins on the bottom of the skirt for three flowers. The flower on the right is 7.5 cm (3 in) up from the hem and 15 cm (6 in) in from the right-hand edge of the skirt front.

STITCHING

The black details on the bees and the blanket-stitch edgings are worked in half thicknesses of wool.

Dress bodice Using green (9018), work long straight stitches for the stems and leaves right across the row. Now add the *tête de boeuf* flower heads in blue (8610), placing the lower flowers over the stems to hold the stems down on the fabric.

Skirt Work the three bluebells in the same way as the bodice.

Pocket With sewing thread, baste around the outline of the bees' bodies then work in stripes of brown (9602) and gold (8156) satin stitch following the shape of the design. Carefully outline the wing shapes in black (9800) back stitch and then, still working in black, add the straight-stitch wing markings, legs and antennae.

MAKING UP

Allow a 1 cm (⅝ in) seam allowance throughout which has been included on the pattern pieces.

Skirt and bodice Place the back bodices and back skirt pieces right sides together and match up the notches. Pin into place and hand sew or machine seam together.

With the front and back pieces right sides together, match up the notches at the shoulders, then pin and sew. With right sides together, and notches matching, pin and sew the side seams of the dress.

Bodice lining Take the bodice lining pieces and lay these right sides together joining notches. Sew together the shoulder seams and then join the side seams. Neaten the bottom edge with a 6 mm (¼ in) seam. Place the lining inside the dress with the wrong sides together. Turn in the raw edges of the calico and lining around the armholes and neckline, baste into position and seam on the right side of the calico as close to the edge as possible. Catch the hem of the lining to waist seam selvedge.

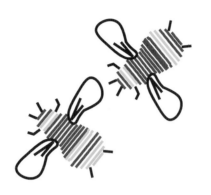

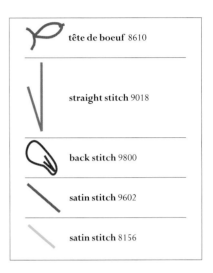

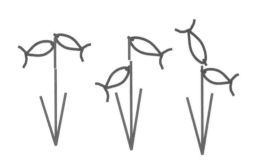	tête de boeuf 8610
	straight stitch 9018
	back stitch 9800
	satin stitch 9602
	satin stitch 8156

Turn in the centre back edges of the lining and calico by 2 cm (1¾ in) and machine seam on the right side of the calico from the top of the neckline to the bottom of the skirt.

Fold up the bottom of the skirt by 6 mm (¼ in) then fold again and baste into place. Secure with a row of hand or machine stitches on the right side of the skirt.

FINISHING

For the back ties, cut four 63 cm (25 in) lengths of lining fabric 5 cm (2 in) wide. Fold in half lengthwise, with right sides together and machine or hand seam. Turn right sides out and hand stitch ends to neaten. Stitch the ties to the back bodice on either side of the back neck opening and slightly above the waistline.

Make a 6 mm (¼ in) hem all around the edges of the pocket and hand or machine seam on the right side of the fabric. Pin into position on the smock and join with blanket stitch in blue (8610) along the curved edge.

Work a continuous row of blanket stitches in blue (8610) around each armhole. Repeat around the neck edge. Press on wrong side over a damp cloth.

Bluebell Mobcap

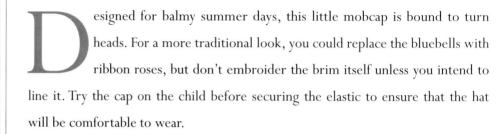

Designed for balmy summer days, this little mobcap is bound to turn heads. For a more traditional look, you could replace the bluebells with ribbon roses, but don't embroider the brim itself unless you intend to line it. Try the cap on the child before securing the elastic to ensure that the hat will be comfortable to wear.

PREPARATION

With the compass and pencil, draw a 25 cm (10 in) diameter circle on the calico. Cut out the circle. Using the water-soluble pen, draw an inner circle measuring 19 cm (7½ in) diameter. Sew a 6 mm (¼ in) hem all around the outer edge. With pins, mark positions for eight flowers, spacing them evenly around the crown of the hat, 2.5 cm (1 in) in from the inner pen line. Place a pin to mark the position of the bee or transfer the bee outline onto the calico.

STITCHING

Work each flower individually. First work the green (9018) straight-stitch leaves and stems. Then work the *tête de boeuf* flower heads in blue (8610) securing the lower flower over the stem. Work the body of the bee in gold (8156) and brown (9602) satin-stitch stripes, then embroider the wings, antennae and feet in black (9800), using only half the thickness of the thread.

MAKING UP

Turn the hat piece over and pin the bottom edge of the bias binding to the inner pen line. Turn to the right side and machine or hand stitch the binding into place along the lower edge. Turn to wrong side and lay the elastic under the binding. Turn over, then pin and stitch down the top edge of the binding. Pull the elastic to the desired tightness and stitch the ends together. Join the short ends of the binding.

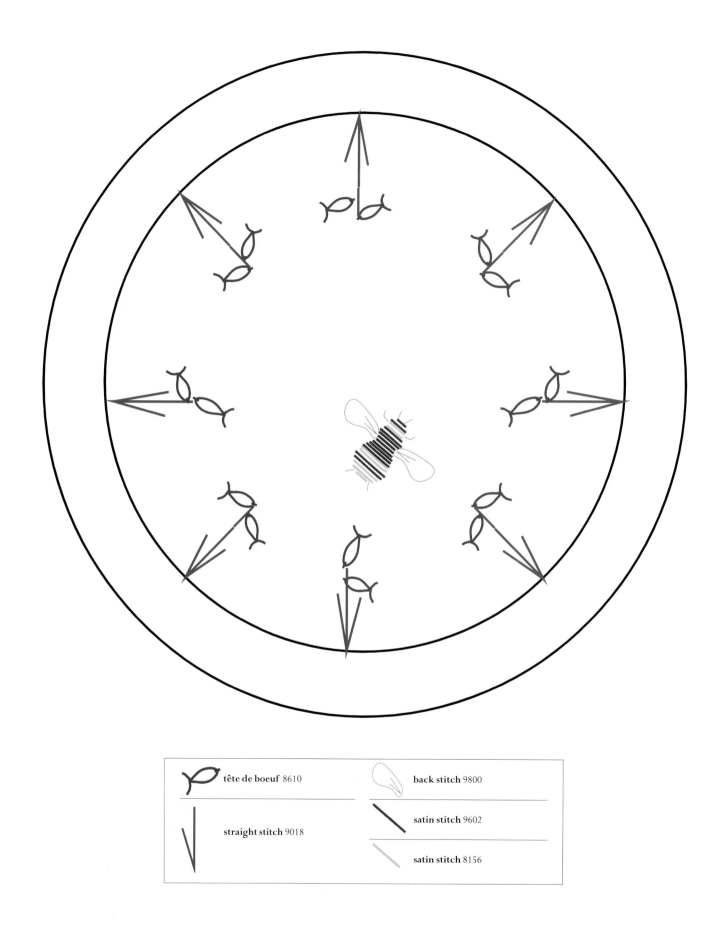

tête de boeuf 8610	back stitch 9800
straight stitch 9018	satin stitch 9602
	satin stitch 8156

Baby Mittens

These little mittens require no seams as they are joined at the edges with blanket stitch. They are really intended for decoration rather than wear and this means you can go to town on the embroidery adding as many flowers and ribbons as you like. They will make the kind of gift that 'baby' might still have around to show his or her children.

PREPARATION

Trace off the mitten outline overleaf and cut two images from blanketing for the left mitten. Turn the template over and cut two more images for the right mitten. Trace off the floral pattern and lay this over one back piece. Push the water-soluble pen through the centre of the main flowers so that the position is marked on the fabric. Turn the tracing upside down and repeat for the second mitten so marking a pattern for a mirror image.

STITCHING

Begin with the straight-stitch roses and using pink ribbon. Next work the bullion roses in pink wool (8139) using two strands. Work the detached chain part flower heads in the colours and yarn indicated on the chart, then add the colonial knots. The small peach cotton (336) flower head at the top left of the chart is produced by working extended colonial knots. Work three, ending the tails in the same spot. Add the fly-stitch leaves and the detached chain-stitch flower heads and straight-stitch stems. Finish by adding colonial knots to fill any gaps in the posy. Make a bow from blue ribbon and catch this into position over the stems.

MAKING UP

Place the back and front pieces of the mittens together with wrong sides facing each other and pin together. Using ecru perlé cotton (926) work in blanket stitch

EMBROIDERY MATERIALS

STITCHES (SEE PAGES 110-113)

Blanket

Colonial knot

Detached chain

Fly

Straight

plus, from the Flower Library
(see pages 114-115):

Bullion rose

Straight-stitch rose

MATERIALS

Pencil

Tracing paper

30 cm (12 in) square of pure wool
blanketing

Water-soluble pen

Medium-sized crewel needle

Pins

YARNS

Silk ribbons

1 m (1 yd) of 6 mm (¼ in) wide in pink

1 m (1 yd) of 8 mm (⅜ in) wide in blue

Anchor Perlé Cotton no. 3

1 skein of each of the
following colours:

Lime 802

Ecru 926

Yellow 290

Peach 336

DMC Medici Crewel Wool

1 skein of Pink 8139 (use two strands)

all around the outside edges so joining the backs to the fronts. Cut two double lengths of ecru perlé measuring approximately 30 cm (12 in) long and twist the two strands together to form a cord and knot one end. Then, using the cord, work a row of large running stitches at wrist level across the back and front beginning and ending at the centre back. Knot the remaining loose ends together, pull to gather slightly and tie with a bow. Repeat for the second mitten.

All stitches are worked in Anchor Perlé unless otherwise described	
	straight-stitch rose 6mm silk ribbon pink
	detached chain stitch 6mm silk ribbon pink
	fly stitch 802
	colonial knot 926
	colonial knot Medici Crewel Wool 8139
	detached chain stitch 926
	bullion rose Medici Crewel Wool 8139
	colonial knot 290
	extended colonial knot 336
	straight stitch 802
	straight stitch 290

Baby Bootees

These little bootees make the ideal christening gift. Here they are made with felt but you could stitch on blanketing or even on kid leather. This pattern will fit a newborn baby, but to adjust the size, lengthen the template. Measure the length of your baby's foot and add 6 mm (¼ in) all around as a seam allowance. While these bootees are made in pink, feel free to change to blue, white, yellow — or, indeed, any other colour that takes your fancy.

PREPARATION

Trace off the template overleaf for the bootee upper and cut out the shape. Fold the felt in half and pin on the template. Draw around the shape with a water-soluble pen and cut out both thicknesses of felt. Do the same with the sole, cutting two shapes in felt and two in blanketing. The embroidered area on the left bootee is a mirror image of that on the right bootee. If you are not confident about copying the design by eye, trace off the stitches and push the water-soluble pen through the centre of each flower so marking the positions on the felt.

STITCHING

First work the straight-stitch rose using white ribbon and then stitch a four-petalled lazy daisy next to it in the fine braid. Above that work a bullion rose in the perlé off white and the adjacent spray of leaves in fly stitch, also in the off white. Work the six-petalled lazy daisy in perlé and the daisy with four petals at the opposite side of the bootee.

Work the two outer sprays of leaves in perlé fly stitch and then work the colonial knots at the top of the bootees. Take the length of white wool and, using half the thickness, work a column of four bell flowers to the side of the bullion rose. Finally, using perlé, work the two daisies, fly-stitch leaves and colonial knots at the back sides of the bootees.

Making up

Place one blanket sole on top of each felt sole. Position the upper on top and pin, then baste into place through the three layers of fabric, around the outer edge and up the back seam. Using perlé, join all seams with a continuous neat row of blanket stitch. Remove basting. Cut the silk ribbon into four equally-sized lengths and pin one length on either side of each bootee. Stitch the buttons into place and, at the same time, secure the ribbon ends under the buttons.

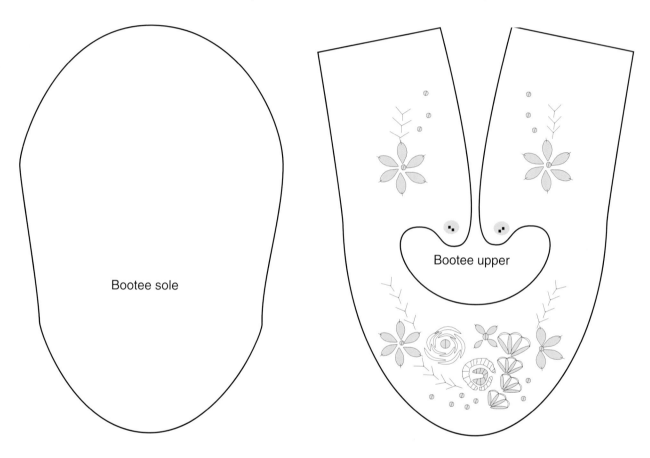

Bootee sole

Bootee upper

All stitches are worked in Anchor Perlé unless otherwise described			
	lazy daisy stitch with colonial knot 2		buttonhole stitch Anchor Tapisserie Wool 8000
	fly stitch 2		straight-stitch rose 4 mm (⅙ in) silk ribbon white
	colonial knot 2		bullion stitch rose 2
	lazy daisy stitch with colonial knot Kreinik fine braid 212		buttons attached using 2

Chicks Dressing Gown

EMBROIDERY MATERIALS

STITCHES (SEE PAGES 110-113)

Bullion

Colonial knot

Fly

Straight

MATERIALS

Pencil

Tissue paper

Pins

Child's towelling dressing gown (terry cloth bathrobe) with pockets

Sewing thread

Medium-sized crewel needle

YARNS

DMC Medici Crewel Wool

1 skein of each of the following colours:

Dark yellow 8725

Light yellow 8027

Orange 8940

Blue 8997

Pink 8151

Green 8420

These little chicklets can be used to decorate children's clothing or cot (crib) blankets. Here they are worked on a shop-bought towelling dressing gown (terry cloth bathrobe): the texture of the towelling helps keep the chicklets' feathers looking fluffy. Simple random straight stitches in two shades of yellow form the body of the design so this is a good project for beginners.

PREPARATION

Trace the outline of the chicklets, overleaf, onto tissue paper and pin one tracing to each pocket front. Baste over the outline with sewing thread and carefully tear away the tissue paper. Do the same with the flower borders on the lapels.

STITCHING

Work the beaks first. These are made with two parallel bullion stitches in orange (8940). Also in orange, work four bullion stitches at angles for the feet and add the blue (8997) colonial knots for the eyes. Thread up some dark yellow (8725) and work rows of random vertical straight stitches to divide the wings and heads from the bodies. Use the black dotted lines on the charts as a guide for position.

Fill in the wings and the tops of the chicks' heads with small random straight stitches in the dark yellow (8725) adding a few strokes of light yellow (8027) on top. Fill in the chests in the same way using mainly the lighter yellow and then adding a few strokes of dark yellow. The flowers are made up of four colonial knots in either blue (8997) or pink (8151) with one centre colonial knot in light yellow (8027). Stitch groups of three or four flowers on either side of the chicklets' feet and then add some random vertical straight stitches among the flowers in green (8420) to represent grass.

The flower borders on the lapels are stitched in the same way as the flowers on the pockets. Work eleven of them in a zigzag row at the edge of each lapel. Join the

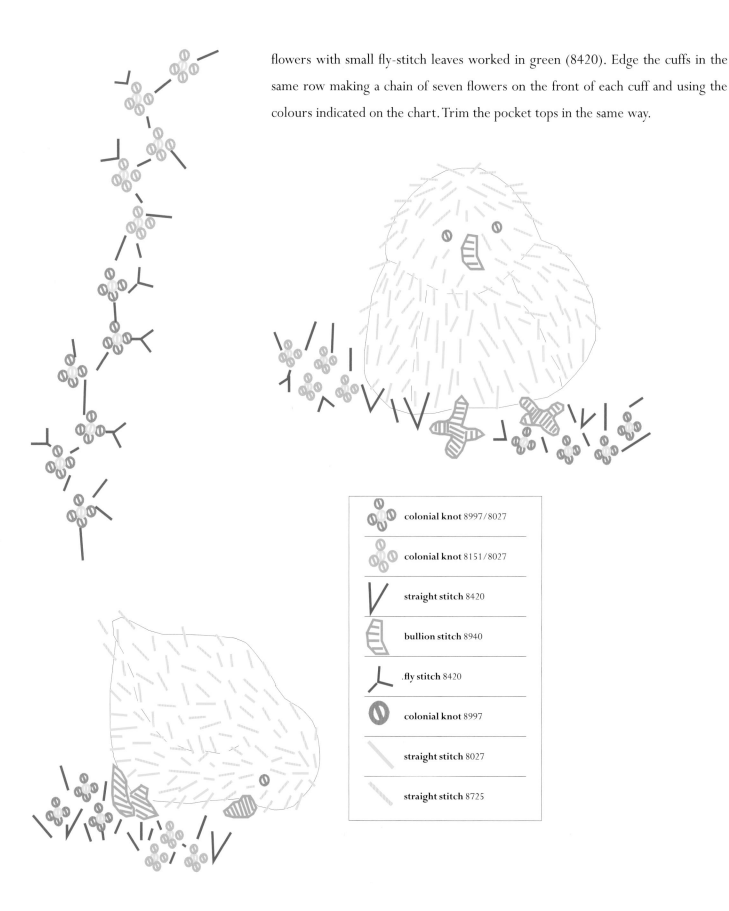

flowers with small fly-stitch leaves worked in green (8420). Edge the cuffs in the same row making a chain of seven flowers on the front of each cuff and using the colours indicated on the chart. Trim the pocket tops in the same way.

	colonial knot 8997/8027
	colonial knot 8151/8027
V	straight stitch 8420
	bullion stitch 8940
Y	.fly stitch 8420
	colonial knot 8997
/	straight stitch 8027
	straight stitch 8725

Teddy Bear Blanket

I had great fun 'painting' these teddy bears with random straight stitches. The Medici wool I have used is very fine which makes it easy to build up the shape by adding darker shades of wool. To make a nice cuddly shape for the head there is one easy trick: point all stitches towards the muzzle and keep the angle of the stitches on either side of the muzzle almost horizontal and the stitches at and above eye level almost vertical.

PREPARATION

Trace off the outline of the teddy bears, overleaf, onto tissue paper. Position the motifs at an angle to the bottom left corner of the blanket, centring the bottom of them approximately 23 cm (9 in) in from the corner point. Using beige (8504) wool and working through the tissue paper and blanket, stitch a firm row of back stitches all around the outlines of the heads, bodies and limbs. Carefully tear away the tissue paper so that just the stitched outlines remain.

STITCHING

Beginning with the standing teddy bear, work the outer ears and right inner paw in beige (8504) straight stitch. Fill in the ears and paw pad in small coffee (8611) random straight stitches. In black (Noir), work the nose and eyes in satin stitch, and then form the mouth from a single fly stitch. Work the muzzle in beige (8504) straight stitches spanning the stitches outwards from the nose/mouth. Still in straight stitches, work the areas surrounding the muzzle and eyes in honey (8322) and the top of the head in beige (8504). Outline the waistcoat in blue (8899) straight stitch and fill in with single vertical straight stitches from top to bottom. Now work single horizontal straight stitches from side to side and stitch down each junction with two crossed straight stitches in yellow (8725). Fill in the legs and arms and with straight-stitch shading, as shown on the chart.

ǀ	satin stitch 8816
✕	straight stitch 8725
▬ ▬	straight stitch 8899
╱	straight stitch 8504
╱	straight stitch 8611
╱	straight stitch 8322
✿	lazy daisy stitch Blanc with **colonial knot** 8725
╱	straight stitch 8344
Y	fly stitch Noir
⬭	satin stitch Noir

The sitting teddy bear is worked in a similar way with the ears and right foot outlined in beige (8504) straight stitch and filled with small straight coffee (8611) stitches. The bow ribbons are worked in pink (8816) vertical satin stitch and the knot in horizontal satin stitch. Follow the chart for the straight-stitch shading on the head, limbs and body.

Make a chain of five-petalled lazy daisies as positioned on the chart and fill each flower with a yellow (8725) colonial knot. Join the flowers with green (8344) straight stitches.

MAKING UP

Press the backing fabric and lay it down wrong side up. Place the blanket right side up on the centre. Making a fine hem at the edge of the fabric, fold in and pin down the two short ends to form a border. Machine or hand stitch these into place close to the edge of the lining fabric. Now turn in and pin down the long ends of fabric and sew down in the same way. Stitch together the open seams at each corner.

Little Teddy Bear

This cuddly little teddy bear is easy to make and perfect for beginners. I have used blue and pink blanketing fabric but, if you have problems finding this, you could make him out of felt. The ribbon attachment is purely for decorative purposes. Do not leave unattended babies with lengths of ribbon as they could wrap it around themselves.

PREPARATION

Trace the template overleaf, cut out the paper shape, pin it to the blue blanketing and carefully cut out the fabric shape. Cut a second body shape from the pink fabric. Then cut four paw pads and two ear pads from the left over pink scraps. Pin paws and ear pads into position on the blue fabric and attach by stitching all around the edges with blanket stitch using white (Blanc) wool. With the water-soluble pen draw two small circles on the face for the eyes and a small triangle for the nose.

STITCHING

Stitch over the eye circles in white (Blanc) chain stitch and fill in the centres with black (Noir) satin stitch. Embroider the nose in pink (8816) satin stitch and then work the mouth using three black (Noir) straight stitches. Work a daisy garland from paw to paw in white (Blanc) lazy daisy stitch and fill in the centre of the flower heads with yellow (8725) colonial knots. Join the flowers with green (8344) straight stitches and then add some pink (8816) colonial knots between the daisies.

MAKING UP

Cut a slightly smaller body shape from the wadding (batting) and sandwich it between the back and front of the teddy bear. Pin the pieces together around the edge. Now join the back and front all around the edges in blanket stitch using white (Blanc).

EMBROIDERY MATERIALS

STITCHES (SEE PAGES 110-113)

Blanket

Chain

Colonial knot

Lazy daisy

Satin

Straight

MATERIALS

Pencil

Tracing paper

Pins

1 piece of pure wool blanketing in blue measuring 23 x 20 cm (9 x 8 in)

1 piece of pure wool blanketing in pink measuring 23cm (9 in) square

Water-soluble pen

Medium-sized crewel needle

Polyester wadding (batting)

YARNS

DMC Medici Crewel Wool

1 skein of each of the following colours:

Yellow 8725

Pink 8816

White Blanc

Green 8344

Black Noir

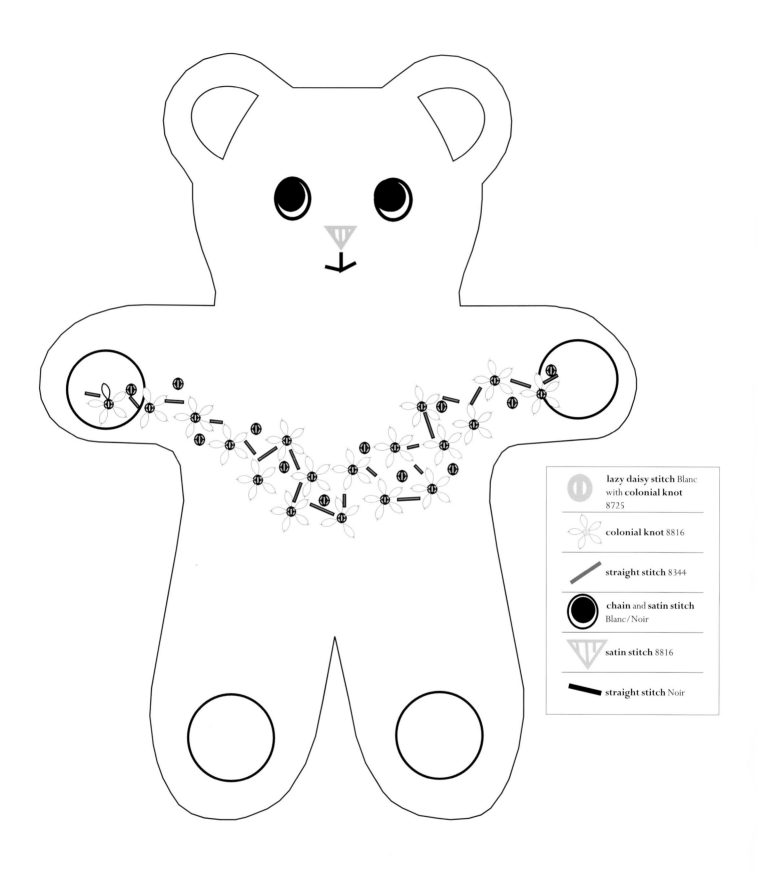

lazy daisy stitch Blanc with **colonial knot** 8725

colonial knot 8816

straight stitch 8344

chain and **satin stitch** Blanc/Noir

satin stitch 8816

straight stitch Noir

Essential Information

Only a few basic materials are required to embroider on wool blanketing and they are easy to get. This chapter explains what you need and why, with an introduction to the basic techniques needed for this craft. A description of how to work all the stitches used in the projects appears in the Stitch Library on pages 110-113, and the Flower Library on pages 114-115 explains how to combine stitches to make the flowers that are most often used in the projects. Finally, on pages 116-122 you will find a number of motifs to help you create your own designs.

Materials

THREADS

Threads for woolwork are available in a variety of textures and thicknesses and also a vast range of colours. The main types of thread are outlined below.

Crewel wools: DMC Medici / Appletons

Sold in skeins, these are very fine single strands of wool that can be doubled up for extra thickness.

Knitting wools

Pure wool yarns can be used very successfully in woolwork. Carefully divide the plies of wool to create the thickness you require. Double knitting wools are particularly useful for edging embroidered throws and other items with blanket stitch.

Persian wools: Paterna / Paternaya

Also available in skeins, Persian wools have a slight sheen and are made up of three separate strands of wool which will divide easily if it is necessary to do this for one of your designs.

Tapestry wools: Anchor / DMC

These are sold in skeins and are designed to be used as a single thickness of thread. However, it is perfectly possible to divide the thread and use half the thickness for delicate work. I have given instructions in the relevant projects in this book where you need to do this.

Perlé cotton

This is a high-sheen, twisted mercerised cotton which is excellent for achieving a textured effect. Normally used as a single strand, cotton perlé is available in various thicknesses from no. 3 (thick) to no. 12 (thin). It is particularly useful for knotted stitches as it shows their raised effect to the fullest advantage.

Alternative threads

Many small spinning and dyeing companies are marketing superb ranges of textured silks, silk ribbons, textured wools and cottons coloured with both natural and synthetic dyes. Freestyle embroidery provides you with the perfect opportunity to experiment with these threads and, used in conjunction with wool, help you to develop your own personal style within your work.

Sewing thread

Always keep a stock of a few reels of sewing thread in contrasting colours. These are used to transfer designs onto your fabric.

FABRICS

You can work wool embroidery on virtually any woollen fabric, but take care to adjust the thickness of your threads according to the weight of the fabric, e.g. use only single strands of crewel wool on lightweight fabric. Also beware of fabrics that are prone to shrinkage and make sure that they are colourfast.

Calico

This most basic of cotton fabrics is ideal for cushions, children's pinafores and tote bags. The combination of crisp calico and soft wools is particularly homely and pleasing. Always pre-wash calico before cutting the pieces for a project as it has an irresistible tendency to shrink.

Felt

Bright, bold images can be worked on brightly coloured handicraft felt which can be purchased in cut squares or by the metre (yard).

Heavy cotton fabrics

Denim and cotton furnishing fabrics such as ticking provide interesting and serviceable backgrounds for wool embroidery.

Knitted fabrics

Jumpers and cardigans can be successfully embroidered with crewel and tapestry wools. Take care not to mix natural and synthetic fibres as the care instructions may not be compatible.

Polar fleece

This is the stuff sport manufacturers use

to line anoraks with. It's thick, it's fluffy and it's a good man-made substitute for wool.

Towelling
Cotton towelling (terry cloth) can be purchased by the metre (yard) in a variety of colours. It is a good washable alternative to wool blanketing.

Wool blanketing
This can be purchased in a range of pastel colours or you can also use travel rugs and cut up existing woollen blankets to provide an interesting variety of background colours. Following the invention of the quilt, old blankets are often unceremoniously disposed of at yard sales, jumble sales and car boot sales where they can be bought at a fraction of their original price. Some blanket manufacturers now mix wool with polyester fibres. Being a 'natural fibre snob' I have requested that you use pure wool blanketing for these projects. This is my preference, but there is no reason at all why it should be yours.

Worsted wools
These are the woollen fabrics you buy to make up skirts and jackets and are fine for wool embroidery. You can also use fabrics made of mixed natural fibres such as cashmere and wool or silk and wool to create shawls and throws.

NEEDLES
It is a good idea to keep a collection of the following needles in your work basket. When selecting a needle for a project, the eye should be large enough to hold the wool and ribbon and the point narrow enough to glide through the fabric with ease. For nearly all of the projects in this book I have used a medium-sized crewel needle without any problems.

Crewel needles
Extremely useful needles with sharp points and large eyes. These are available in sizes 1-10 and are perfect for woolwork.

Chenille needles
Chenille needles are thicker and longer than crewel needles and they have larger eyes and sharp points. These are ideal for thicker threads and are available in sizes 14-26.

Tapestry needles
Blunt ended, large-eyed needles available in sizes 14-26. Used in wool embroidery for lacing and whipping threads.

MISCELLANEOUS ITEMS
Dressmaker's chalk pencil
Used for positioning designs on the blanketing. See transferring designs, overleaf.

Scissors
You ideally need three pairs of sharp scissors: one large pair of dressmaking shears for cutting fabric; one pair of all-purpose scissors for cutting out paper patterns, and a pair of small embroidery scissors for snipping thread ends.

Tracing paper and tissue paper
This is used for tracing patterns and transferring designs to your fabric.

Transfer pencil
See transferring designs, overleaf.

Water-soluble pen
A pen like this allows you to draw directly onto your fabric. Once you have finished your stitching, sponge the pen marks with cold water and they will disappear.

TRIMS
Look out and hold on to the following, all of which can be useful to the woolwork embroiderer:
wide ribbons, bias bindings and furnishing braids and fringes are useful for binding the edges of blankets. Fabric prints and sheeting can be used for lining blankets.
Buttons, beads and charms make interesting additions to 'flower borders'. It is not, however, advisable to use these on baby items as they could be pulled off and swallowed.

Techniques

FRAMES

One of the great pleasures of embroidering on wool blanketing is that it is totally unnecessary and undesirable to use a frame. You can sit on a couch on a cold winter's evening and drape the blanketing over you as you sew. The major peril is cat and dog hair but I'm sure your animals are too well behaved to 'deliberately' initial your work in this way! When working on calico or heavyweight cotton fabric, a frame is probably desirable although I did not use one when stitching the designs in this book. A circular frame that is slightly larger than the finished design would be your best choice.

TRANSFERRING DESIGNS

When working on blanketing your best method of transferring an outline such as a teddy bear is as follows:

Method 1

Trace the outline from the template onto tissue paper and pin the tracing into position on the fabric. With a fine needle and a sewing thread that contrasts with the fabric, work a row of running stitches through both paper and fabric all around the outline of the motif. Tear away the tissue paper and use the running stitches as guidelines, removing the thread when your stitching is complete.

Method 2

Trace the outline from the template onto lightweight card and carefully cut around the shape. Pin this into position on your fabric and draw around the outline with a dressmaker's chalk pencil or a water-soluble pen.

For a floral border, trace the design onto tissue paper and pin this into position on your fabric. Pierce the paper at the centre of each flower with a pin and then mark through the hole with a water-soluble pen. Use the resulting dots as guidelines to position your flowers.

When working on calico and cotton fabrics, any of the above methods can be used. Alternatively, trace the design onto tracing paper using a pencil. By turning over the paper to the wrong side and drawing over the pencil lines with a transfer pencil you now have a reverse image. Place the completed tracing transfer-side down on your fabric. Press a medium-hot iron onto the tracing and count to five. Lift the iron, taking care not to slide it as you do so, and then the paper and the image should have transferred onto the fabric. It is best to test this method on a scrap of the fabric before attempting to transfer the complete design. You might find that you need to keep the iron on the paper for a little longer.

Commercially printed transfers

These can be ironed onto blanketing fabric but some people find the process a little tricky. Heat the iron to the wool temperature and press it flat over the reverse of the transfer for a count of six. Then remove, taking care not to slide the iron.

ENLARGING AND REDUCING DESIGNS

The simplest way of achieving enlarged and reduced designs is on a photocopier. Most modern copiers give you a number of size differentials to choose from. Where appropriate, the project instructions in this book tell you by what percentage you need to increase a design.

The traditional method is to trace your design onto squared paper and then to redraw it, carefully replicating the image in each squared section using a larger or smaller scale of squared paper depending on the size of design you are trying to achieve. But do try to find a photocopier – the squared method is labour intensive.

CREATING MIRROR IMAGES

The simplest way to create a mirror image is to trace off the design onto tissue paper, tracing paper or clear acetate using a felt-tipped pen. Turn the tracing over and the wrong side provides you with your mirror image.

RIBBON TIPS

Threading the needle

Silk ribbons have been used in several projects and this method of needle threading will ensure you get the most from a length.

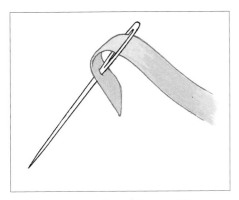

1 Using a crewel needle push the ribbon end through the eye and pierce this end of the ribbon with the needle point.

Pull the ribbon through and then gently slide the pierced end up the needle so that it is anchored securely at the top.

Making ribbon leaves (right)

Using the basic straight stitch you can taper the end of the ribbon to create a simple leaf shape.

1 Bring the ribbon up at A. At the point where you would normally go down, B, pierce the ribbon with the needle and take the needle down through the fabric.

2 Pull the ribbon through gently so that it forms a folded point.

Fastening on

When starting a new piece of ribbon, leave a small tail at the back of your work and carefully secure this with subsequent stitches.

Fastening off

Weave the end piece of the ribbon in and out of adjacent stitches at the back of the work before finally cutting off the tail end.

FINISHING EDGES

Although blanketing is not prone to

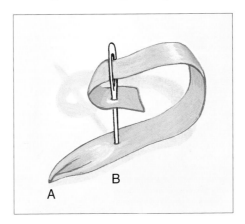

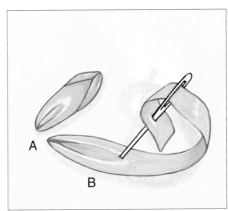

excess fraying, a neatly finished edge will add to the quality of your design. Blanket or buttonhole stitch is the obvious choice, but, when working a large piece such as a knee rug, you might consider binding the edges with ribbon or backing the design with cotton (see Teddy Bear Blanket on pages 98-100) and folding over the edges to form a contrasting hem. An outline of running stitch a short distance in from the edge of your blanketing will also break up large areas of blank space and help to balance your composition and create an interesting border. You can also deliberately fray the edges of the blanketing by easing away the threads with a needle. Do not fray more than ten threads or the fibres will matt together when washed. There are also numerous furnishing trims and laces that you can buy to edge the wool.

AFTERCARE

Don't machine wash woollen blanketing. Treat it with the care your time and effort deserves and wash it by hand in luke warm water and use soap. Any serious stains can be dealt with individually when the blanket is dry. After washing, gently wring out the excess moisture and lay it over an airer to dry. When pressing, do so with the embroidery face down over a fluffy towel to avoid flattening the fibres.

Stitch Library

The following step-by-step diagrams give you the method of creating all the stitches used in the projects in this book. In addition to the stitches featured here, you will also come across **straight stitch** very frequently in these projects. Also known as stroke stitch, it is the most basic of stitches as it is worked by coming up where you wish the stitch to begin, and down where you wish it to end.

Back Stitch

Use this stitch to outline your motifs and create stems for your flowers.

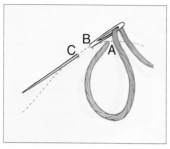

1 Come up at A, down at B and up at C.

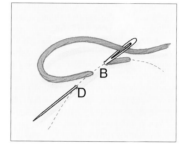

2 Take the needle back and down at B, coming up a stitch length away at D.

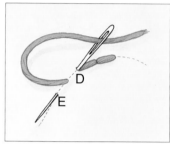

3 Continue the row going back and down at D and up at E.

Blanket Stitch, Buttonhole Stitch

This is the stitch that works so successfully as a decorative binding for fabric edges. Its success depends on the even spacing of the stitches. In buttonhole stitch, the stitches are placed flush to each other, in blanket stitch there is a space between each stitch.

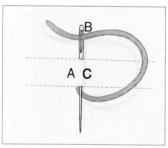

1 Come up at A, down at B and up at C, looping the thread under the point of the needle.

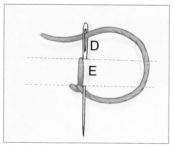

2 Go down at D and up at E, looping the thread under the needle.

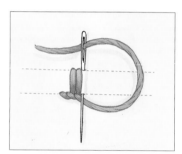

3 Continue in this way, keeping the top of the stitches in an even line.

Bullion Stitch

This stitch can be worked individually to form a corded straight stitch or in a group to form a flower head or bullion rose (see the Flower Library, page 114).

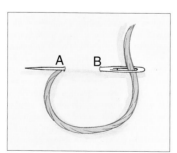

1 Come up at A, go down at B and come up at A again to form a horizontal stitch.

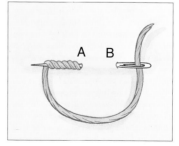

2 Wrap the thread around the needle point until the distance between A and B is covered.

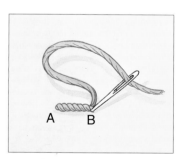

3 Hold the coil together with your thumb and forefinger and pull the needle through the coil. Then go down again at B to secure.

Chain Stitch, Detached Chain Stitch and Lazy Daisy Stitch

An individual chain stitch is known as detached chain and is used for small leaves. Worked in a circle, detached chain becomes lazy daisy stitch which is perfect for flower heads. If you work a continuous row of chains you are stitching basic chain stitch.

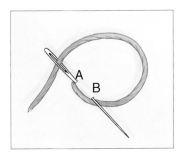

1 Bring the needle up at A, down directly beside it and come up at B, looping the thread under the point of the needle.

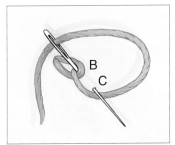

2 For a detached chain, secure the loop with a straight stitch, or to continue, go down beside B, and come up at C, looping the thread under the needle.

3 Lazy daisy stitch consists of five or six detached chain stitches worked in a circle.

Colonial Knot

This useful little knot is a simple alternative to the traditional French knot. It can be used alone to represent buds or flower centres, or you can group four together to create forget-me-nots. An **extended colonial knot** is worked in the same way, but to finish, insert the needle a short distance away from the knot to form a straight stem.

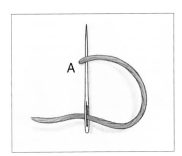

1 Come up at A and lay the needle on the fabric to the right of A.

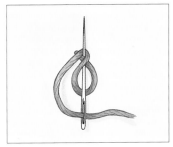

2 Carefully wrap the thread around the needle to form a figure eight.

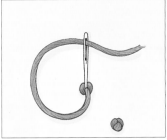

3 Then pull the thread through and go down in the centre of the knot as illustrated.

Cretan Stitch

This simple looped stitch is used for making leaves. Draw guidelines on your fabric before beginning.

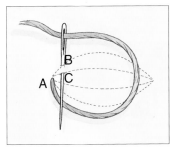

1 Come up at A, go down at B and come up at C looping the thread under the point of the needle.

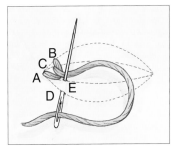

2 Go down at D and come up at E. Once again, loop the thread under the needle.

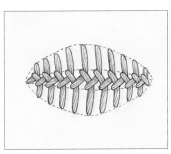

3 Repeat steps 1 and 2 until your leaf shape is complete.

Feather Stitch

This very traditional and simple looped stitch is lovely for fern-like sprays of leaves and also for flower stems.

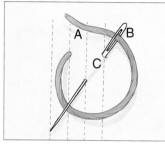

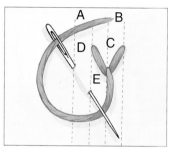

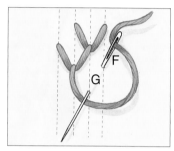

1 Come up at A, go down at B and then come up at C, keeping the thread under the needle point.

2 Go down at D and come up at E with the thread under the needle point.

3 Continue as set to form a straight or curved column of stitches, depending on what your requirements are.

Fly Stitch

Another useful little stitch for both leaves and flower petals. It can be worked individually or in a column.

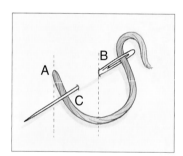

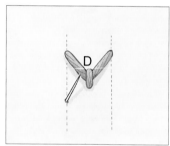

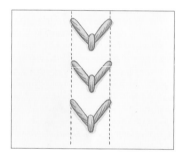

1 Come up at A, go down at B and up at C laying the thread under the needle point.

2 Go down again at D to secure the stitch.

3 Work this stitch individually or in a vertical column, as shown, depending on your requirements for the design.

Satin Stitch

Use this stitch when you want a solid filling for a motif such as a flower bud. Take care to graduate the edges evenly.

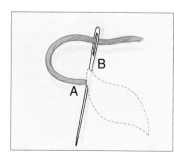

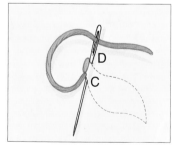

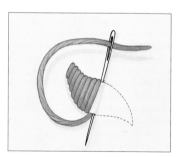

1 Come up at the edge of your motif at A and down at the opposite edge B.

2 Come up again at C (flush to A) and then down at D (flush to B).

3 Continue until your motif is filled with stitches.

Split Stitch

This is another useful stitch for creating a fine, even outline. It can also be used for stems and stamens.

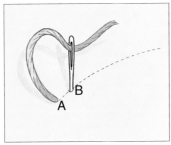

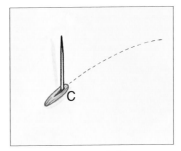

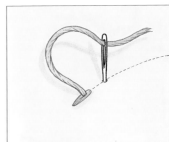

1 Come up at A and go down at B to form a small and neat straight stitch.

2 Come up at C, halfway between A and B, piercing the thread of the existing stitch.

3 Continue in this way carefully following the line of your design.

Stem Stitch

Here is another useful stitch for creating stems and curved lines. Always bring your needle up above the previous stitch rather than below.

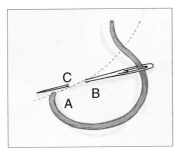

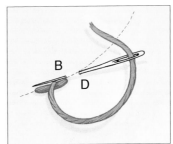

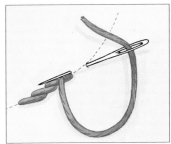

1 Come up at A, go down at B and come up again at C (halfway between A and B).

2 Go down at D, so forming a stitch equal in length to the first stitch. Come up again at B.

3 Continue in this way carefully following the line of the design.

Tête de Boeuf Stitch

While this stitch is supposed to resemble the 'head of a bull', it is also perfect for making bell-shaped flower heads.

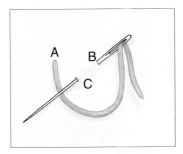

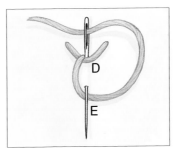

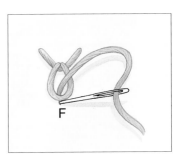

1 Come up at A and go down at B forming a horizontal stitch. Come up again at C.

2 Make a detached chain stitch by going down at D and coming up at E, looping the thread under the needle point.

3 Go down at F in order to secure the loop.

Flower Library

Many stitches resemble flowers or parts of flowers, and by combining them cleverly you can fill an embroidery with perennial favourites. By carefully selecting colours, threads, stem lengths, and leaf shapes, you can produce bold or delicate floral arrangements using the same basic stitches.

Bell Flowers

This is a tall plant with bell-shaped flowers and can be used to represent delphiniums, marshmallows, foxgloves or any other flower with these characteristics. Master the blanket stitch (see page 110) and then have a go at working a bell flower.

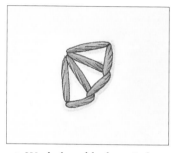

1 Work three blanket stitches from one central point.

2 Repeat step 1 but change the angle of the flower and position it above the first flower.

3 Repeat steps 1 and 2, then join the flowers with a back stitch stem and work two detached chain stitch 'leaves' at the base.

Bullion Rose

This flower is created from bullion stitch (see page 110). Each stitch represents a rose petal, and the more stitches you add, the larger the rose.

1 Work a bullion stitch beginning and ending at the same point so that the stitch forms a small circle.

2 Make two more stitches, curving them around the central stitch and crossing over the ends so the stitches are slightly raised.

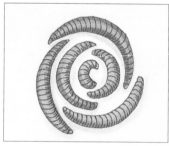

3 Continue to add stitches as in step 2 until the rose is the required size.

Hollyhocks

Large columns of hollyhocks are very picturesque and particularly effective if you graduate the shades of the flowers. Each flower consists of buttonhole stitch (see page 110) worked in a circle to form a 'wheel'. Work a column of flowers on a central stem reducing the size of each wheel towards the top.

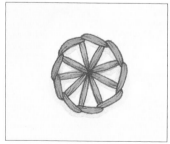

1 Make a circle of buttonhole stitches working each stitch from a central point.

2 Continue building a column of buttonhole circles making each one smaller than the last.

3 Make a stalk by joining the flowers with a row of back stitches and finish with a one-third buttonhole circle in green to form a leaf at the stem base.

Honeysuckle

These wonderful flowers can be stitched in wool or ribbon in various shades of yellows, peaches and pinks. The main petals are 'wrapped', which gives the effect of raising them slightly off the fabric.

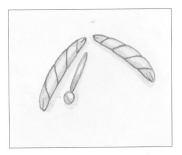

1 Make a straight stitch from A to B and come up at C. Take the needle over and under the stitch, repeating until the stitch is covered.

2 Make a second wrapped stitch, positioned as shown. Then stitch a colonial knot, extending the finishing stitch to form a stamen.

3 Repeat step 2 so that you have a cluster of stamens of differing lengths between the two petals.

Spider's Web Rose

This is my favourite rose to work with ribbon. According to the size of the rose, you can create a base of three or five spokes. You can also turn the ribbon slightly as you weave it to get a more raised effect.

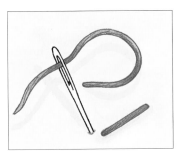

1 With ribbon or wool, make a fly stitch, positioning the spokes evenly.

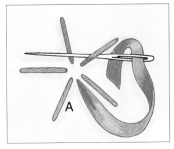

2 Come up at A and take the ribbon over the first spoke and then slide it under the second spoke without going through the fabric.

3 Continue weaving in and out of the spokes until they are completely covered.

Straight-Stitch Rose

This stitch combination works in both ribbon and wool and is a simple, effective way of creating a rose. For added effect you can shade the colours of the petals keeping the centre quite dark to give the flower depth.

1 Work three small straight stitches to form a triangle. Then work three stitches around the outside of the triangle.

2 Continue in this way until you have the size of rose you require. Do not pull the thread too tightly so that the petals appear curved.

3 For a variation, use different colours of ribbon for each layer of petals.

Motif Library

Among these pages you will find motifs that you can trace off and use on your own choice of projects. Some of these motifs are simple outline shapes which can be filled in with straight stitches and others indicate the stitches that should be used.

When filling in areas with straight stitch it is usually a good idea to split your wool in half and to work with single strands. This technique allows you to build up the texture of the image. Avoid pulling the threads too tightly or your fabric will pucker. Wool blanketing in particular has a tendency to stretch, so it is wise to keep your stitch strokes light and handle your work with care.

Colours

When selecting yarns to work these designs, you have a number of choices to make. Shades of a single colour throughout a design can be extremely effective and, if this is an idea that appeals to you, work some of the flowers in cotton perlé or silk ribbon to give the image depth and shine. When choosing shades it is a good idea to vary the tones only slightly and then introduce one stark contrasting tone to bring the composition to life.

As an alternative, you can, of course, use rich contrasting colours but keep the tones within the same family. Purples, reds and golds look wonderful with the addition of metallic threads and rich silk ribbons.

Embellishments

Wool flowers are also shown to great effect when they appear in a dense cluster. Once you have completed a motif, fill any gaps with extra fly stitch leaves, buttons, beads or colonial knots. Small glass beads also make perfect flower centres. The easiest way of applying these is with ordinary sewing thread or stranded cotton in a shade that is slightly darker than the bead itself. Gilt charms and chains also make a pretty addition to your work.

Arranging motifs

The motifs can be used on their own for small projects or they can be repeated or mirrored to form a border. Take a number of photocopies of your motif and arrange them on your fabric until you have a composition you are pleased with. You can also enlarge or reduce the images and adjust the size of your stitching accordingly.

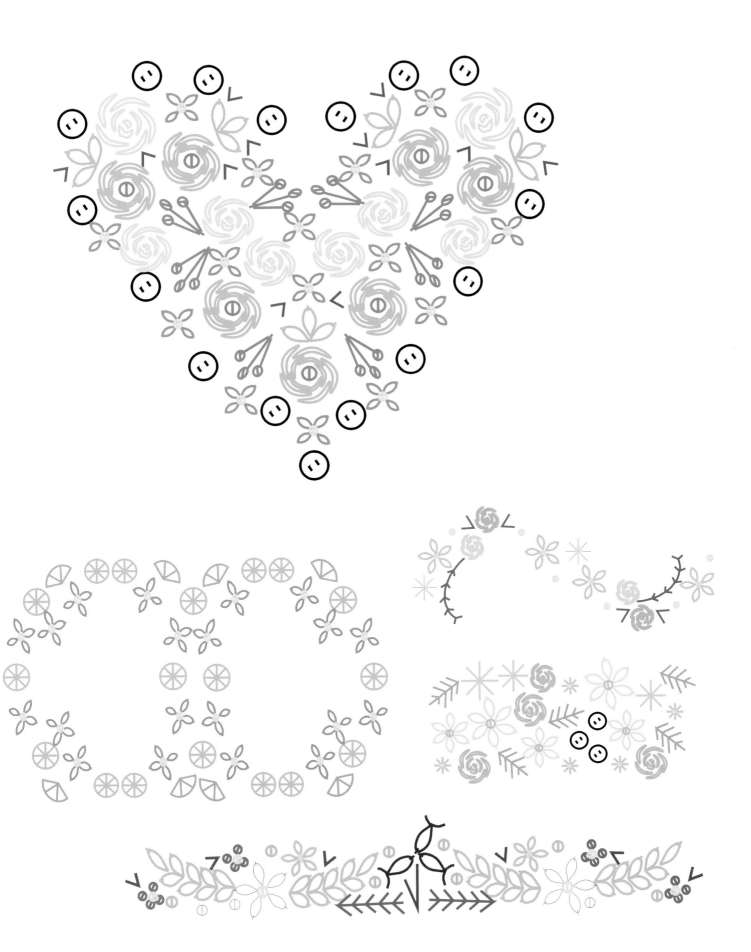

Templates

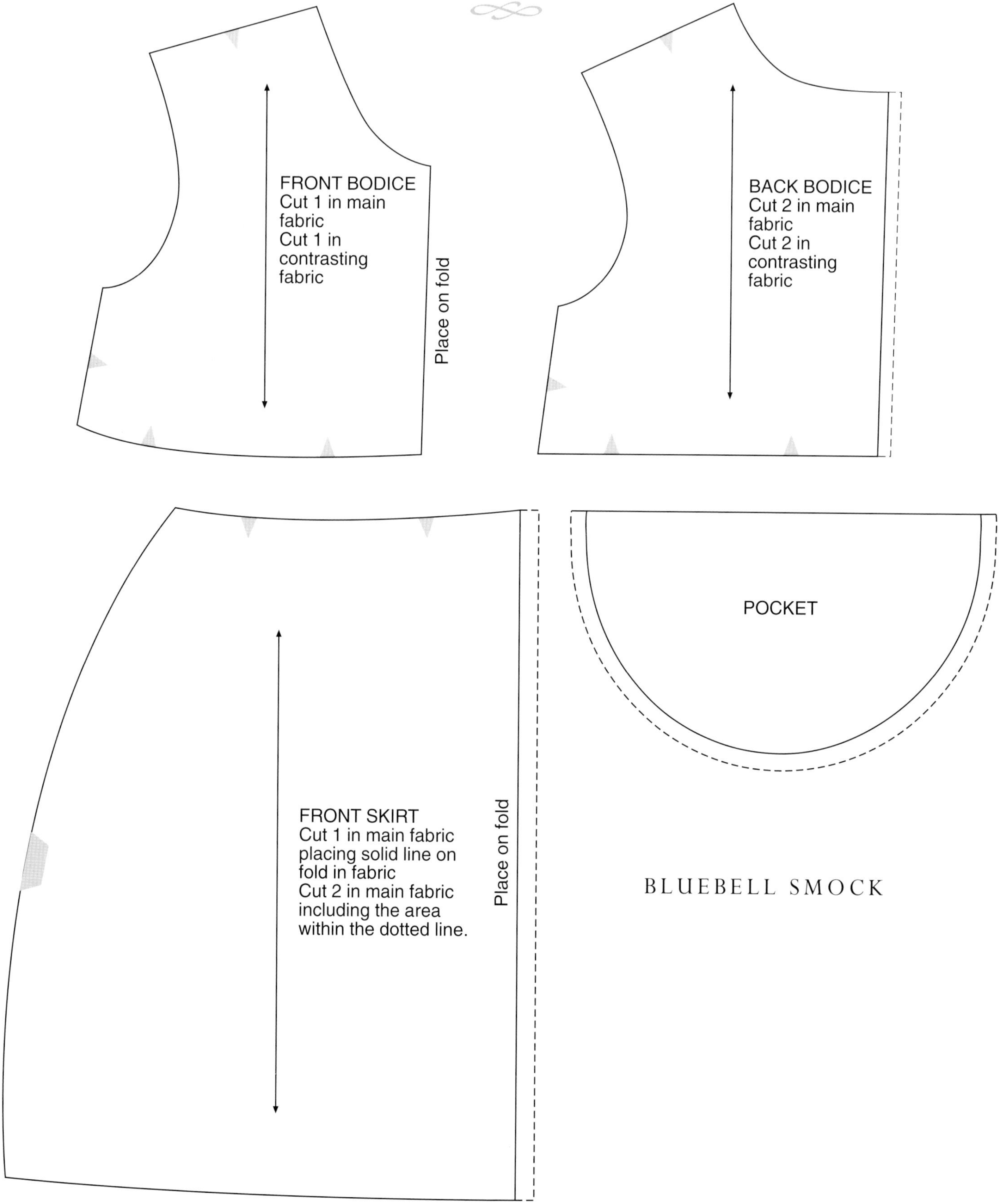

FRONT BODICE
Cut 1 in main
fabric
Cut 1 in
contrasting
fabric

Place on fold

BACK BODICE
Cut 2 in main
fabric
Cut 2 in
contrasting
fabric

FRONT SKIRT
Cut 1 in main fabric
placing solid line on
fold in fabric
Cut 2 in main fabric
including the area
within the dotted line.

Place on fold

POCKET

BLUEBELL SMOCK

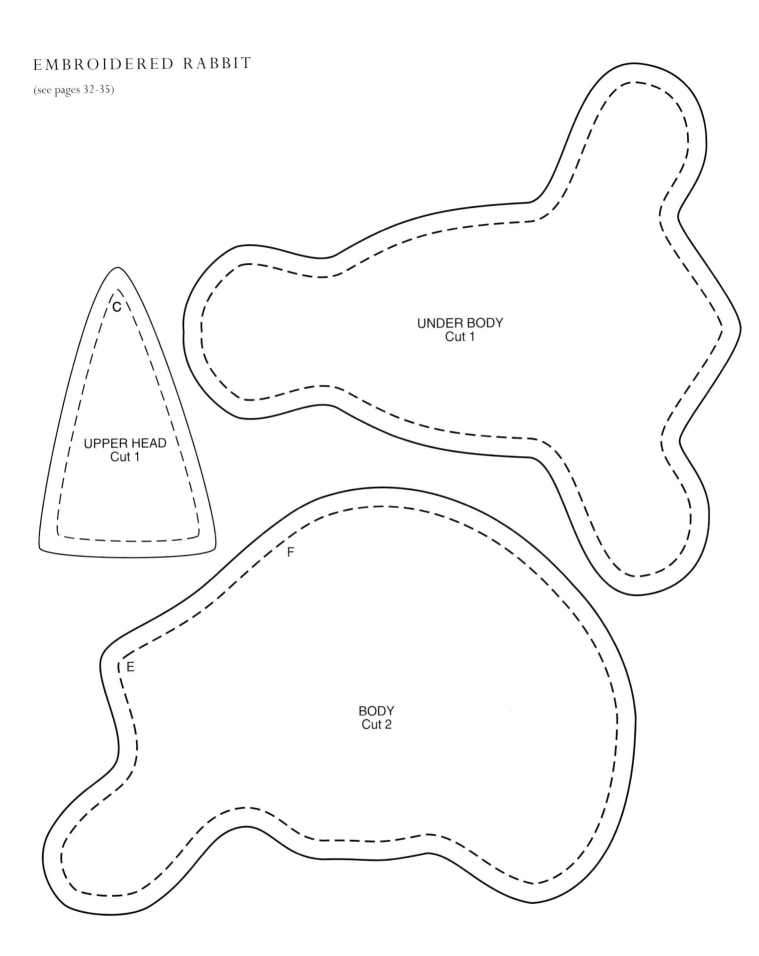

UPPER HEAD
Cut 1

C

UNDER BODY
Cut 1

F

E

BODY
Cut 2

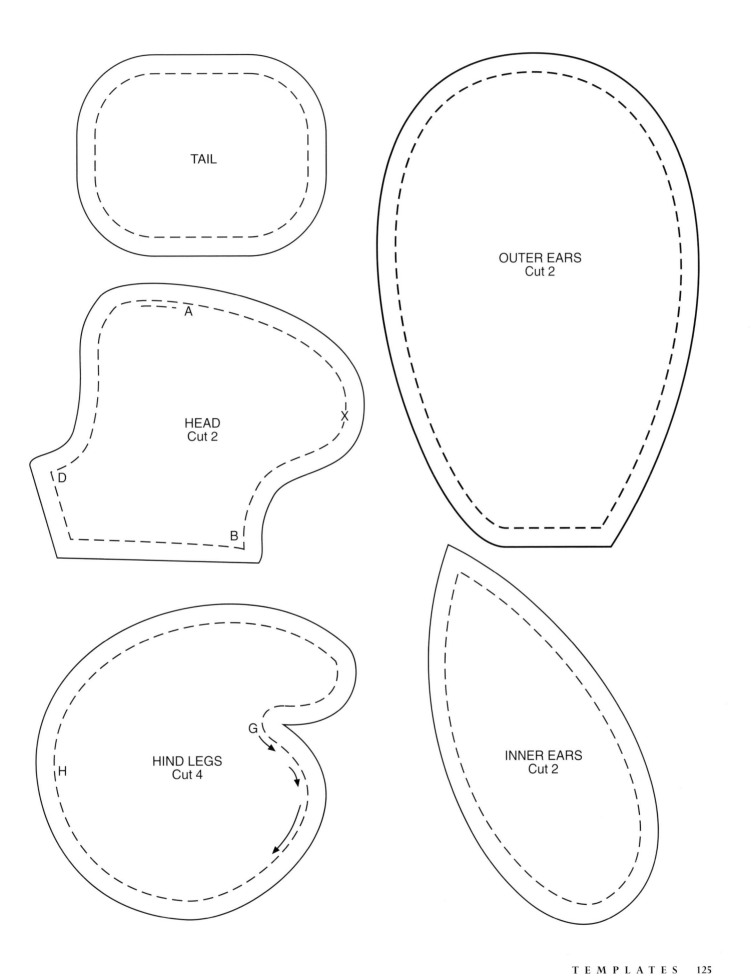

TAIL

OUTER EARS
Cut 2

HEAD
Cut 2

A

X

D

B

HIND LEGS
Cut 4

G

H

INNER EARS
Cut 2

Stockists and Suppliers

Blanketing

The beauty of wool blankets is that they are readily available in department stores throughout the world. The following suppliers also offer a mail order service or will provide you with a list of local stockists:

Wool blanketing and hand-dyed silk ribbons are available by mail from:

Melinda Coss, Ty'r Waun Bach, Gwernogle, Dyfed, West Wales SA32 7RY
Telephone and fax: 01267 202 386 (office hours only please)
email mcoss@cix compulink.co.uk

For local stockists of Anchor Tapisserie Wools, Cotton perlé and metallic threads, write to:

Coats Patons Crafts, 382 Wellington Road, Mulgrave, Victoria 3170, Australia

Coats FFR NV, Burgemeester de Cocklaan 4, 9320 Erembodegem-Aalst, Belgium

Coats Sartel Loisirs, 22 rue de la Tannerie, BP 39, 59392 Wattrelos Cedex, France

Coats Mez GmbH, Postfach 1179, D79337, Kenzingen, Germany

Coats Cucirini Spa, Via Vespucci, 2, 20124 Milano, Italy

Coats Multicem A/S, Fjoesangerveien 50, Posttboks 1135, N 5001 Bergen, Norway

CA Linha Coats & Clark, LDA, Coats Crafts Marketing, Quinta de Cravel, 4430 Vila Nova de Gaia, Portugal

Coats Fabra SA, Sant Adria 20, 08030, Barcelona, Spain

Coats Molnlycke Sytrad A/S, Kallbacksrydsgatan 8, Box 245, S - 501 05 Boras 1, Sweden

Coats Crafts UK, PO Box 22, Lingfield Estate, McMullen Road, Darlington, Co. Durham, DL1 1YQ, United Kingdom

Coats & Clark Inc., 30 Patewood Drive, Suite 351, Greenville, S. Carolina 29615, USA

In addition, the following company stock all metallic yarns produced by Coats:

Kreinik Mfg. Co., Inc., 9199 Reisterstown Rd, Suite 209B, Owings Mills, MD 21117, USA

For local stockists of Medici crewel wool write to:

DMC Needlecraft Pty Ltd, 51-66 Carrington Rd, Marrickwille NSW 2204, Australia

DMC, 7/9 Rue du Pavillion, B-1210 Brussels, Belgium

DMC, Dampfaergevej 8, Frhavnen, DK-2100 Copenhagen, Denmark

DMC, 10 avenue Ledru-Rollin, 75579 Paris, Cedex 12, France

DMC, Viale Italia 84, I-20020 Lainate Mi, Italy

DMC, Travessa da Escola, Araujo 36-A, P-1100 Lisbonne, Portugal

DMC, Fontanella 21-23, (Floors 4 & 5), E-08010 Barcelona, Spain

DMC Morgenstrasse 1, CH 9242, Oberuzwil SG, Switzerland

DMC Creative World Ltd, Pullman Road Wigston, Leics LE18 2DY, United Kingdom

The DMC Corporation, Port Kearny, Building 10, South Kearny, NJ 07032, USA

Acknowledgments

Thanks are due to the following creatures for their continued support and confidence in my work:

My assistants Carol Jones and Jeanette Hall who have, in addition to their studio support, taken care of the important things in life like coffee making, feeding the piglets and taking the cats to the vet.

My editors: Cindy Richards and Emma Callery who are probably sick to the teeth of me by now.

Mary Coleman and the team at *Woman's Weekly Handmade* whose wonderful Australian magazine inspired this book.

Sue Stravs at *Embroidery and Cross Stitch*, ditto.

Coats Crafts UK and DMC for their beautiful supply of wools.

Jillian Taylor of Needle Needs for her lovely harvest mouse design.

Jon and Barbara Stewart for their superb photography and also Roger Daniels for his book design.

Mark Collins and Cameron Brown for having the courage of my convictions.

Index